THE ILLUSTRATED
Wild Flower Finder's CALENDAR

THE ILLUSTRATED

Wild Flower Finder's CALENDAR

David Lang

Bounty Books

First published in Great Britain in 1983 by
Ebury Press under the title
The Wild Flower Finder's Calendar

Published in 1985 by Chancellor Press, an imprint of Bounty Books,
a division of Octopus Publishing Group Ltd

Reprinted 2001

This edition published in 2012 by Bounty Books,
a division of Octopus Publishing Group Ltd
Endeavour House
189 Shaftesbury Avenue
London WC2H 8JY

www.octopusbooks.co.uk

An Hachette UK Company
www.hachette.co.uk

ISBN: 978-0-753723-54-8

A CIP catalogue record for this book is available from the British Library

Printed and bound in China

Illustrations by Stuart Lafford, Steve Lings, Jim Channell, Tim Hayward and Clive Spong
by courtesy of Linden Artists Ltd
Designed by Richard Garratt Design
Concieved and edited by Neil Curtis Publishing Services

Contents

The pleasures of finding wild flowers

As a member of my school cadet force, I was involved one day in a field exercise on Ashdown Forest in Sussex. Never outstanding for my military ability, I fell full length in a boggy patch of ground, to find my eyes on a level with masses of glorious blue gentians. Full of excitement, I was certain that I had found the Trumpet Gentian of the Alps *(Gentiana acaulis)*, only to be subdued by a friend who had greater botanical knowledge and informed me that it was Marsh Gentian *(G. pneumonanthe)*, The memory of those flowers has never left me. Years later I revisited the spot – the gentian was still there and, in rediscovering it, the pleasure was as fresh as ever.

Finding a particular wild flower for the first time in your life is an exciting experience. If it is a rare species then the pleasure is enhanced but there is as much enjoyment to be obtained from the sight and scent of well-loved flowers encountered freshly each year. The luck of discovery need not include measuring your length in the mud and the excitement is perhaps even greater when a particular species is found after long and diligent searching. I shall never forget finding my first Lady Orchid *(Orchis purpurea)* in a Kentish wood. In the bare earth beneath the yew trees, the stately flower spikes, with broad, shiny leaves, seemed to glow in the dim light. Closer examination of the flowers only added to the delight, each flower resembling a little figure in a crinoline, marked with minute tufts of mauve hairs.

Looking for flowers in the mountains adds another dimension to hill walking, scrambling, and climbing. Mountain flowers are often small and delicate, tucked away in sheltered crannies in the rocks, on ledges, and in damp crevices. The first Starry Saxifrage *(Saxifraga stellaris)* I found was growing in a runnel of sparkling water on a scree slope below the summit of Cader Idris in north Wales. Although I have found the species many times since then, for it is not uncommon in Wales, the Pennines, and Scotland, the joy of that first discovery remains. One of the pleasures of a holiday in a new place is the difference in the wild flowers you find. As a southerner I shall never forget the first sight of Bird's-eye Primrose *(Primula farinosa)*

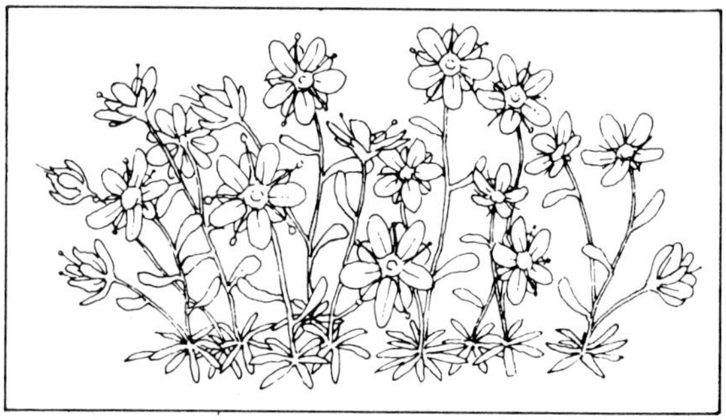

Yellow Saxifrage

growing in profusion above Wharfedale in Yorkshire early on a June morning. Yellow Saxifrage *(Saxifraga aizoides)* is widespread in limestone areas in the north of Ireland, northern England, and especially north-west Scotland, but, to someone unaccustomed to seeing it, the carpet of yellow flowers beside mountain streams, or making neat yellow cushions on damp rocks is a source of delight.

Many of our wild flowers have insignificant blooms compared with the large, brightly coloured flowers of cultivation or growing wild in warmer countries, yet close examination will reveal delicate colours and patterns. The tiny flowers of Eyebright *(Euphrasia officinalis)* are of great beauty, despite their small size and lack of bold colour, although you really need a magnifying glass to appreciate them. It is a common plant in grassy places all over Britain, and is particularly fine and brightly coloured on the machair of the Hebrides. Eyebright is also a good example of a species with many close relatives which show subtle variations in colour and structure. The study of such flowers is an absorbing and rewarding hobby, adding to the knowledge and understanding of our flora; it need not involve long journeys or strenuous activity, and can be done by anyone with a little time to spare and the willingness to learn.

There is great satisfaction in finding a wild flower in an area where it has never been recorded, or where it has only been seen on a few occasions. Pipewort *(Eriocaulon aquaticum)* is a rare plant of freshwater lochs in Skye, Coll, Mull, and western Ireland. In western Argyll it had long been known in two sites and, during a walking holiday there, we found it in another six by searching every remote loch we came upon. In several of them it was flowering abundantly. The rosettes of leaves can pass unnoticed as they lie a metre or more below the surface, until the plant flowers, sending up a long stem crowned with a white flower head like a small button above the water. Since then it has been found in several more lochs so that its known range has been considerably increased. It is not possible to comprehend the size and remoteness of the mountainous country of north-west Scotland unless you spend time walking in it. There could be other exciting discoveries yet to be made there by those with sharp eyes and a love of the hills.

For most of us who enjoy looking for wild flowers, there will be elusive plants that we persistently search for, but fail to find, even though they are not necessarily rare. Such a one for me was Cyphel *(Cherleria sedoides)*, a cushion-forming plant with tiny yellow-green flowers, which grows at high altitude on bare mountain tops in north-west Scotland. It gave me great pleasure finding the neat green cushions for the first time in Sutherland, even though the site was a well-known one, and the discomforts of tiredness, cold, and driving rain were overcome.

Not all plant hunting need involve strenuous walking or climbing. If you can find copies of old floras and plant lists, especially those which describe the flowers to be found in your home area, you can have an enjoyable time checking these old records to see if the plants are still there. I well remember finding Moschatel *(Adoxa moschatellina)* and Toothwort *(Lathraea squamaria)* still flourishing in a Kentish copse more than 100 years after they had originally been recorded there. It is not essential to live in or visit the wild places to enjoy flower hunting. People who live in urban or industrial areas can have a very amusing time with 'aliens' and 'adventives'; these are plants which do not properly belong to the British flora but have been brought here artificially. Hooked and spiny seeds travel in imported wool and cotton, while others came in grains and animal feeds, so that many unexpected places can yield the most extraordinary and bizarre flowers, from docks where the material is landed, to waste tips where the spoil is dumped after processing.

Above all, the study of wild flowers is fun. They are always interesting and beautiful, while the search for them will take you to some of the loveliest parts of Britain.

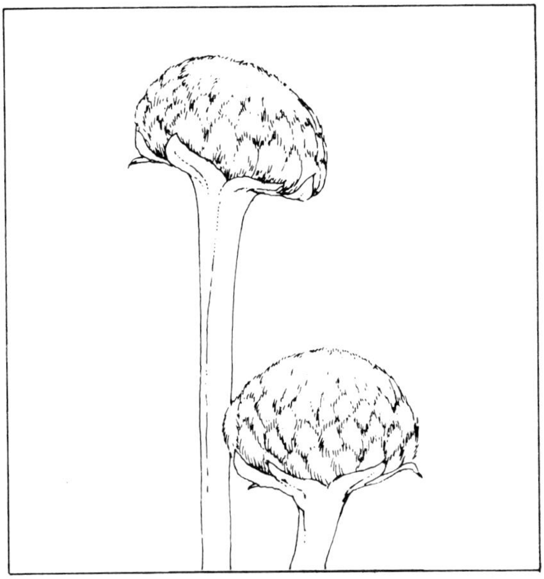

Pipewort

Naming and classifying wild flowers

Plants have always played a vital role in our history, both as sources of food and as medicines for ourselves and our domesticated animals. Inevitably names were given to plants and flowers, names in the common speech of the area which reflected the uses to which they could be put, or names based on a fanciful resemblance of the plant to some everyday object.

By the time printing had been developed, plants had become more widely used by herbalists so that many old books contain vernacular names which are not easily recognizable today. Latin was the language of the medical and scientific world, and efforts were made to identify plants and flowers using Latin descriptions. Although these attempted to describe clearly the nature of the plant, they were often long-winded and cumbersome, such as *Orchis bulbis indivisis, nectarii labio quinquefido punctis scabro, cornu obtuso, petalis distinctis* to describe the Military Orchid.

Military Orchid

Carl Linnaeus was born in 1707 in Småland, a province in the south-east of Sweden. From his clergyman father, a keen gardener, he learned the names of all the common plants. During his studies at Uppsala he became friendly with Petrus Artedi, a student from northern Sweden, and together the two young men pledged themselves to record all plants and animals in the world in a methodical manner. Artedi was drowned in 1735, leaving Linnaeus to continue the monumental task on his own. He undertook journeys to Lapland and to the Baltic islands of Öland and Gotland at the request of the Swedish parliament, to record the plants, customs, and agricultural methods of the people, with an eye to their profitable development. In the index to that report he first used the system of describing a plant by two Latin names. This he developed into a comprehensive binomial system, and the Linnaean system of classification, first published in the *Genera Plantarum* of 1753, is the basis of modern taxonomy.

There are obvious advantages in describing plants using Latin, because botanists from any country, whatever language they speak, will know what plant is meant by the Latin or scientific names. Over time, the process has been refined, and the methods of plant classification have become internationally accepted but, in that process, the scientific names which were first used to describe some plants have been altered, as our understanding of the relationships between plants has improved. Each plant has two scientific names: the first, or generic name, is common to that plant and to those other plants which are obviously so similar to it in structure that they are placed together in a group called a genus; the second name is the specific or trivial name, and applies to that plant only. The generic name usually has an initial capital letter whereas the specific name does not; both are printed in italic type. Thus, the Daisy has the scientific names *Bellis perennis,* belonging to the genus *Bellis* and the species *Bellis perennis.* In printed text it is usual to print the scientific names fully when they are first mentioned, but to abbreviate the generic name for further mentions in that piece of text, for example, *B. perennis.*

When making a full reference to a plant, the name of the author (the first person to distinguish the species) is printed after the scientific name. In the case of Linnaeus this is abbreviated to L.; thus, Bulbous Buttercup is *Ranunculus bulbosus* L. Where the original scientific name has been changed, as the result of our improved knowledge of the species and its position within a related group of plants, the name of the original author appears in brackets, followed by the name of the new author. Greater Butterfly-orchid was originally called *Habenaria chlorantha* by Custer and subsequently assigned to the genus *Platanthera* by Reichenbach. So *Habenaria chlorantha* Custer becomes *Platanthera chlorantha* (Custer) Reichenbach. The whole plant kingdom is divided into branches by a system of classification based on structure, and the divisions are then arranged in the order in which they are assumed to have evolved, from the primitive to the complex. This may be expressed in a simplified form as follows:

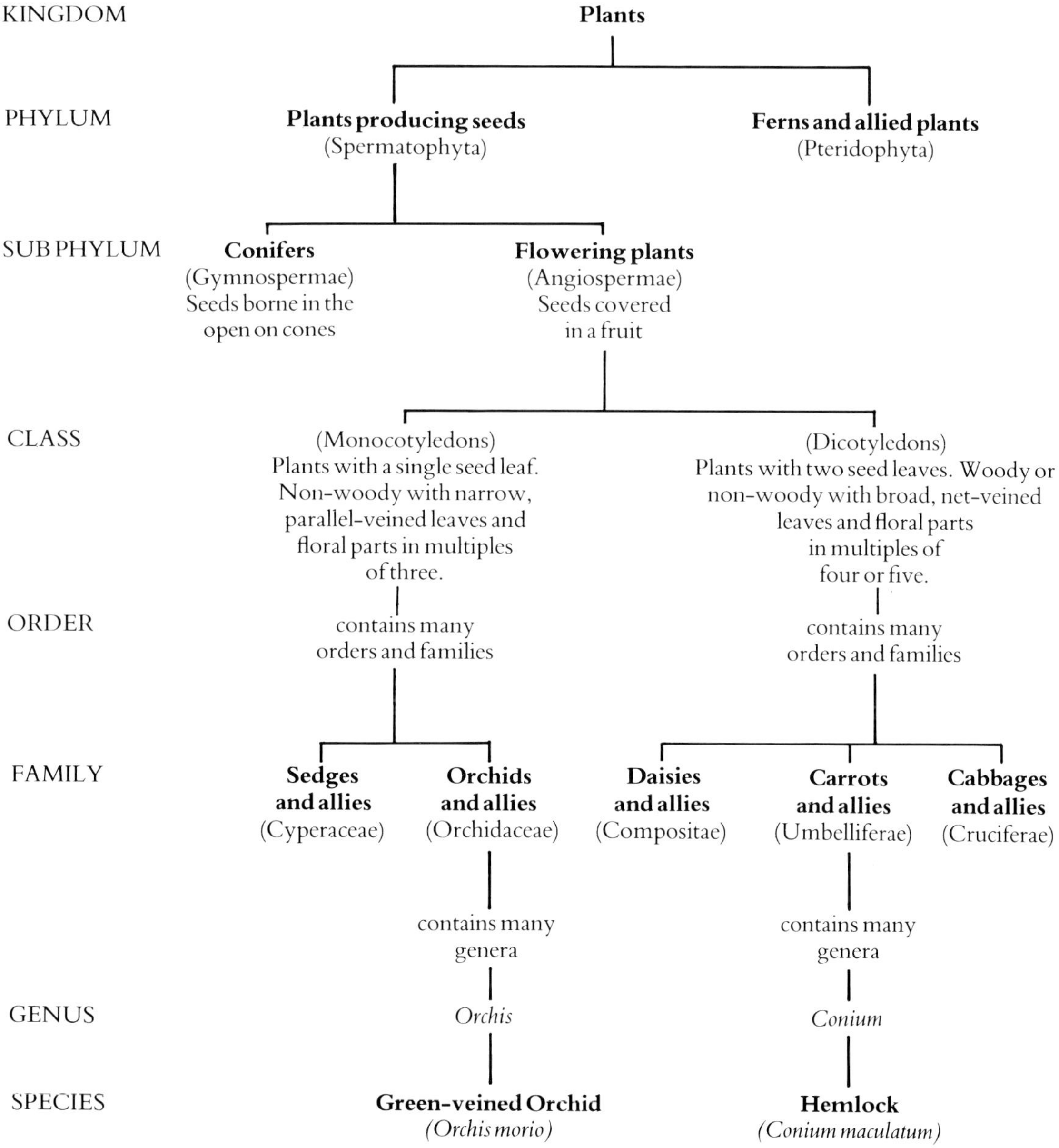

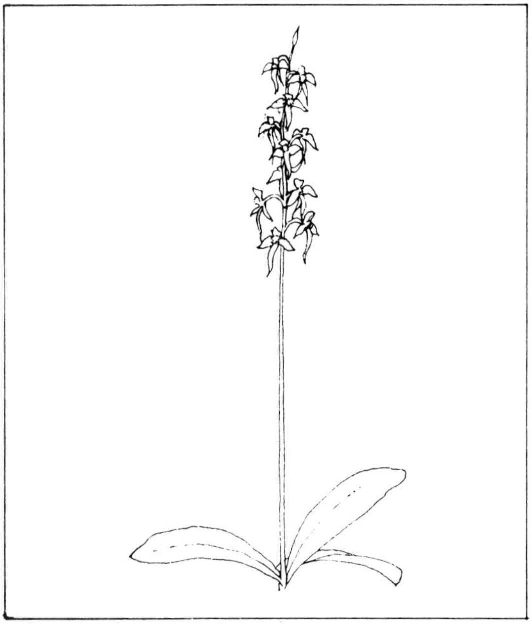

Greater Butterfly-orchid

The scientific naming of plants is well organized and its usage widely accepted but the same is not true of the common English names. Plants often have many different names in use in different parts of Britain, and one English name may even be applied to completely unrelated plants. For this reason the Botanical Society of the British Isles asked three eminent botanists, John Dony, Franklyn Perring, and Catherine Rob, to recommend one English name for each wild plant, to produce order out of chaos. *English Names of Wild Flowers* was published in 1974, and I have adhered to the system set out in that book for the English names used in this volume.

It would be a sad thing to lose entirely the quaint or local names of our wild flowers, such as 'Jack-go-to-bed-at-noon' for Goat's-beard *(Tragopogon pratense)* whose flowers open in the morning and close in the afternoon, but there will be less confusion if standard English names are used in all publications. The value of scientific names becomes especially obvious if we consider problem groups such as the lady's-mantles *(Alchemilla),* hawkweeds *(Hieracium),* and eyebrights *(Euphrasia).* Many of the numerous members in each group have no common English name, and their scientific names are essential.

In describing hybrids it is customary to write the scientific names of the parent plants in alphabetical order. Thus the hybrid between the Common Spotted-orchid and the Southern Marsh-orchid is written *Dactylorhiza fuchsii* x *praetermissa,* while the hybrid between the Common Spotted-orchid and the Frog Orchid, which is of a different genus, is written *Coeloglossum viride* x *Dactylorhiza fuchsii.*

The study and conservation of wild flowers

Few people who have an interest in natural history in any of its branches can be unaware of the pressures and conflicts of interest which threaten much of our wildlife. Many of the problems are inevitable in a country as small and densely populated as ours, so it is a good thing to know about the various organizations which are concerned with nature conservation, and to see where our own interests and activities fit into the general pattern.

The Nature Conservancy Council (NCC) is the government agency responsible through its chairman to the Secretary of State for all matters relating to nature conservation. The Conservancy is staffed by full-time officers, with regional offices which are concerned with the maintenance of National Nature Reserves (NNRs) and who also advise on Sites of Special Scientific Interest (SSSIs), which they certify to government departments, statutory bodies such as water authorities, local authorities, and the owners of the land involved. SSSIs need not necessarily have reserve status, but contain certain plants or wildlife of particular value and interest, the welfare of which must be supervised. The NCC also maintains centralized teams of specialists such as the Chief Scientist's Team, who operate through all the regions. Local Nature Reserves (LNRs) are the concern of the local authorities, who create and manage them.

Most counties in Britain have their own county naturalists' trust, or trust for nature conservation. These are amateur organizations maintained by the subscriptions of their members, and dedicated to the conservation of wildlife within their county borders. Many of these trusts run their own reserves, either owning them or having management rights. The activities of the county trusts are co-ordinated by the Royal Society for Nature Conservation.

Natural history societies, schools, and other organizations also own or manage smaller local reserves and, although all these bodies are independent, there is considerable co-operation, at local and national level, for their mutual benefit and for the benefit of conservation generally.

All those interested in wild flowers should actively support their own county naturalists' trust, as well as local societies catering for their specific interests. Apart from their job of creating and managing reserves, the trusts organize lectures and guided outings for their members, which are of particular value to novices in the study of wild flowers and for people who are new to an area. Most trusts also organize groups within the county to cater for local needs. For those wishing to take an active part in improving the environment for flowers and animals, many trusts run a conservation corps, where volunteers work on a variety of schemes such as scrub clearance, tree planting, reed cutting, or drainage control, under the guidance of experts. If you have difficulty in finding the address of your county naturalists' trust, your local library will be able to help. One of the most important facets of the work of the trusts is the collection and co-ordination of data on wildlife. Records of wild flowers of particular interest may be sent to the Botanical Recorder or Scientific Officer for the county (or part of the county if the area is particularly large), who can be contacted through the trust. This feedback of field information from members is vitally important in maintaining an up-to-date picture of the wild flowers within the county, especially where species are endangered. Many of the county floras are out of date or out of print, so that the trusts are involved in gathering information to compile new floras; botanical recording groups may be formed especially for that purpose. It is essential that such information is kept fully up-to-date, otherwise it is impossible to organize conservation management on a rational basis, or to be aware of areas which may be threatened by some alteration in their usage. Data on rare species will always be treated confidentially – it is no good complaining about the loss of a rare flower if, to protect it, no-one has been told of its presence.

If you become really absorbed in the study of wild flowers, I would strongly recommend applying to join the Botanical Society of the British Isles. The BSBI operates throughout Britain, with meetings, lectures, conferences, outings under the guidance of experts, and visits abroad to areas of outstanding botanical interest, led by members of the BSBI and regularly publicized. It is a lively society, which provides a wealth of fascinating information, not only in an excellent newsletter, but with more erudite papers in the magazine *Watsonia*. Enquiries should be addressed to the Hon. General Secretary, BSBI, c/o Botany Department, British Museum (Natural History), Cromwell Road, London SW7 5BD. The BSBI appoints botanical recorders for each county to whom plant records for that area can be sent, and regular lists of new county records appear in *Watsonia*.

The growth of interest in wild flowers is very heartening, but it brings with it an inevitable increase in pressure on the countryside and the very plants we should be striving to protect. Those interested in wild flowers are often puzzled and upset that they should be kept out of areas where rare and interesting plants grow; they feel that the good things are being reserved for a privileged few. The problem of allowing people access in large numbers to areas of fragile habitat is a constant worry. The basic difficulty is erosion of habitat caused by foot traffic and, even the best-intentioned people can inadvertently destroy young seedling plants. It is essential in some cases to prohibit all visits to an area so that the habitat may have time to recover. In addition, botanists passing to and from plant sites create an obvious trail, and many rarities are still picked and uprooted by selfish and unscrupulous 'flower lovers'. Photography has to a large extent obviated the need to pick wild flowers, and a good photograph is, in most cases, preferable to a discoloured and desiccated pressed flower. Photographers should resist the temptation to 'garden' too rigorously around their subject and should restore things to their normal positions if they are moved temporarily. When it is necessary to remove a flower or a leaf for study, this should be done with care, and the minimum should be taken. Where a flower is obviously rare, or exists as a single specimen, then it should not be touched, and the old guiding rule of taking the book to the flower and not the flower to the book should prevail.

With the passing of the Wildlife and Countryside Act (1981), sixty-two wild flowers have become fully protected. They may not be picked or dug up without a special permit, which may be granted only under exceptional cirucumstances, and even an attempt to do so is an offence which carries a penalty fine of up to £500. The species fully protected by the Act are listed at the end of this chapter. In addition to the sixty-two special wild flowers, it is an offence to uproot intentionally *any* plant without the permission of the owner or occupier of the land, or their agent. Never enter private property in the search for wild flowers without first trying to secure the permission of the owner. Most farmers and land owners will prove amenable to a polite request for access, and it is the responsibility of the visiting naturalist to observe the courtesies of the countryside, avoiding damage to crops and ensuring that all gates are securely closed so that stock cannot stray.

It is well worthwhile keeping a wild flower diary, or some other more detailed record of the wild flowers which capture your interest. Using a local flora with the excellent Ordnance Survey maps opens up a new and fascinating world. There are few parts of Britain which do not have places with wild flowers that are interesting to visit. Should you find something new, a plant never before recorded in that area, make careful notes, including a detailed map reference, so that the site is easy to find again, and make sure that the information reaches the right authority. Some of the most exciting finds have been made by visitors on holiday in remote areas or places seldom visited by botanists. Only if the organizations are informed can they take steps to safeguard rare plants, and this is a process in which we all have a part to play.

List of plant species fully protected under the Wildlife and Countryside Act 1981

Killarney Fern	*Trichomanes speciosum*
Dickie's Bladderfern	*Cystopteris dickieana*
Oblong Woodsia	*Woodsia ilvensis*
Alpine Woodsia	*W. alpina*
Adder's-tongue Spearwort	*Ranunculus ophioglossifolius*
Small Alison	*Alyssum alyssoides*
Fen Violet	*Viola persicifolia*
Alpine Catchfly	*Lychnis alpina*
Cheddar Pink	*Dianthus gratianopolitanus*
Childling Pink	*Petrorhagia nanteuilii*
Teesdale Sandwort	*Minuartia stricta*
Norwegian Sandwort	*Arenaria norvegica*
Perennial Knawel	*Scleranthus perennis*
Rough Marsh-mallow	*Althaea hirsuta*
Rock Cinquefoil	*Potentilla rupestris*
Wild Cotoneaster	*Cotoneaster integerrimus*
Plymouth Pear	*Pyrus cordata*
Drooping Saxifrage	*Saxifraga cernua*
Tufted Saxifrage	*S. cespitosa*
Field Eryngo	*Eryngium campestre*
Small Hare's-ear	*Bupleurum baldense*
Sickle-leaved Hare's-ear	*B. falcatum*
Purple Spurge	*Euphorbia peplis*
Sea Knotgrass	*Polygonum maritimum*
Blue Heath	*Phyllodoce caerulea*
Diapensia	*Diapensia lapponica*
Sea Lavender	*Limonium paradoxum*
Sea Lavender	*L. recurvum*
Spring Gentian	*Gentiana verna*
Alpine Gentian	*G. nivalis*
Spiked Speedwell	*Veronica spicata*
Greater Yellow-rattle	*Rhinanthus serotinus*
Field Cow-wheat	*Melampyrum arvense*
Bedstraw Broomrape	*Orobanche caryophyllacea*
Thistle Broomrape	*O. reticulata*
Oxtongue Broomrape	*O. loricata*
Wood Calamint	*Calamintha sylvatica*
Downy Woundwort	*Stachys germanica*
Limestone Woundwort	*S. alpina*
Water Germander	*Teucrium scordium*
Jersey Cudweed	*Gnaphalium luteoalbum*
Field Wormwood	*Artemesia campestris*
Least Lettuce	*Lactuca saligna*
Alpine Sow-thistle	*Cicerbita alpina*
Ribbon-leaved Water-plantain	*Alisma gramineum*
Starfruit	*Damasonium alisma*
Whorled Solomon's-seal	*Polygonatum verticillatum*
Snowdon Lily	*Lloydia serotina*
Round-headed Leek	*Allium sphaerocephalon*
Wild Gladiolus	*Gladiolus illyricus*
Lady's-slipper	*Cypripedium calceolus*
Red Helleborine	*Cephalanthera rubra*
Ghost Orchid	*Epipogium aphyllum*
Fen Orchid	*Liparis loeselii*
Late Spider-orchid	*Ophrys fuciflora*
Early Spider-orchid	*O. sphegodes*
Lizard Orchid	*Himantoglossum hircinum*
Military Orchid	*Orchis militaris*
Monkey Orchid	*O. simia*
Triangular Club-rush	*Scirpus triquetrus*
Brown Galingale	*Cyperus fuscus*
Starved Wood-sedge	*Carex depauperata*

How to use the book

The wild flower enthusiast in Britain is well served by a wealth of books which cover many aspects of the subject. County floras and floral atlases map the distribution of plants thoughout Britain; field guides describe the flowers in detail so that they can be identified even by a novice; and there are guides to areas of special interest, describing the wild flowers which grow there.

For the beginner, however, it is not easy to discover what can be found in flower in any particular month of the year. It is to fill that need that *The Illustrated Wild Flower Finder's Calendar* was conceived. It covers England, Scotland, Wales and Ireland.

The main section is divided as a calendar, with an introduction which sets the flower scene for each month, drawing particular attention to places where things of interest are happening. Two-hundred-and-twenty wild flowers are illustrated and described in detail. The species are placed in the month when they are usually at their best although they may well flower during other months of the year. The time of year at which a plant flowers will depend on the altitude of the habitat and on how far north it is. Primroses *(Primula vulgaris)* usually flower in March in Sussex and in May in Sutherland while Bluebells, *(Endymion non-scriptus)* which are fully out in April in woods in the south may still be in flower in late June on north-facing seacliffs in Scotland.

In each month the flowers are arranged in the same order in which they appear in the *Atlas of the British Flora,* which is the order also adopted in most of the floras in current use. Ferns, mosses and liverworts, grasses and sedges, trees, and shrubs are not included in this book.

Beside each illustration the height range of the plant, if erect, is given in millimetres (eg. 50-80 mm) or, if the plant is very tall in metres (eg 1-1·5 m). Where the plant grows in a flat cushion or creeps on the ground, instead of height, the width of the flower is given after an asterisk (eg ★3 mm) so that a clear idea of size can be obtained.

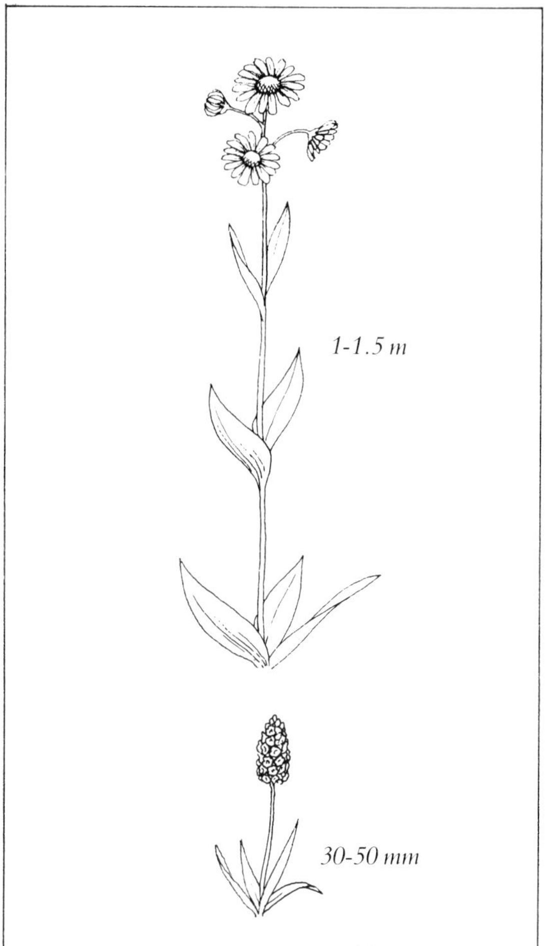

Nearly 5000 species of wild plants have been recorded growing in Britain and, from these, 220 have been carefully selected. Many of these are common and widely distributed throughout Britain so that the beginner can find them easily. Others are less common, and the search for them will take you to many different types of habitat such as seaside, woods, and mountains. A few are rare and are of special interest, so that they present a challenge to the wild flower lover who must spend time and trouble to find them. The flowers have been selected to give as wide a coverage as possible of the different wild flower families, although with 220 examples it is not possible to be completely comprehensive.

During the winter months very few plants can be found in flower although the unusual can still surprise and delight us, while in spring and summer the abundance of wild flowers is such that it will take many years and much travelling to see all that this country has to offer. On pages 25 to 35 the different types of plant habitat are described, including examples of some of the flowers which grow there; and the distribution of these various habitats in Britain is discussed.

At the end of the book the yearly cycle of each of the 220 flowers selected is illustrated in a chart. The flowers are arranged in alphabetical order of their common names so that they can be found more easily. The chart gives a guide to the activity of each plant in every month of the year, whether it is just visible above the ground, in flower, or in fruit; it also gives a reference to the types of habitat in which the plant can be found.

No botanical expertise is required to use the book. For any time of the year, consult the introduction to that month, which will tell you the general areas of interest at that time. The selection of flowers for the month will give you a guide to some of those you will see in flower, and the text will describe where they grow. Using the common names, check the chart at the end of the book, which will also show the type of habitats. Reference to the chapter on habitat will guide you to the areas where that habitat exists, and tell you of some of the associated species you will find there.

The Illustrated Wild Flower Finder's Calendar is best used in conjunction with a field guide or excursion flora although it can be used on its own. They are designed to enable you to identify any flower you find by pictures, colour coding, or keys using botanical structure. On page 180 you will find a list of the field guides and floras currently available, and from them you can select those which suit your taste and interests. There is much to be said for having several field guides, for each will have its particular strong points. The *Atlas of the British Flora* contains distribution maps for all wild flowers, trees, shrubs, and ferns in Britain. Many counties have their own county flora which details the distribution of all the plants within the county boundary. They are essential for anyone who wants to study the wild flowers in a given area, and are full of fascinating information on the present and past status of the flowers.

Although the countryside has changed so much in the last fifty years, the old floras and plant lists can provide immense interest and fun. It is fascinating to play botanical detective, and to search for flowers in places where they were recorded many years ago. It is a source of wonder that something as delicate and vulnerable as a wild flower can be found flourishing in the same wood, stream, or mountain gully where it was described over 100 years ago.

Another pleasant diversion, for those with an artistic streak, is to colour in the black-and-white drawings in the old Bentham and Hooker *British Flora* of each wild flower as you find it for the first time. You can also use the more recent *New Illustrated British Flora* by Butcher. This not only enhances the appeal of the plates, but helps you to understand the structure of the flower, and you will remember it more easily in the future.

Equipment

Field equipment can be simple, consisting of a stoutly covered notebook, a sharp penknife, and a few screw-topped plastic tubes in which small specimens can be collected, as well as the ever-useful polythene bag. To study difficult and variable species you will need to build up a reference collection and individual petals, leaves, or other specimens can be mounted under transparent adhesive tape on cards. These can be kept indefinitely, repeatedly handled without damage, sent through the post to other interested persons, and stored in a very small space. A small hand lens with a magnification of x8 or x10 is an essential piece of equipment. With it you will be able to see details of structure to enable you to identify difficult species.

With this book, a field guide, a flora, and these basic tools, the world of wild flowers is open to you.

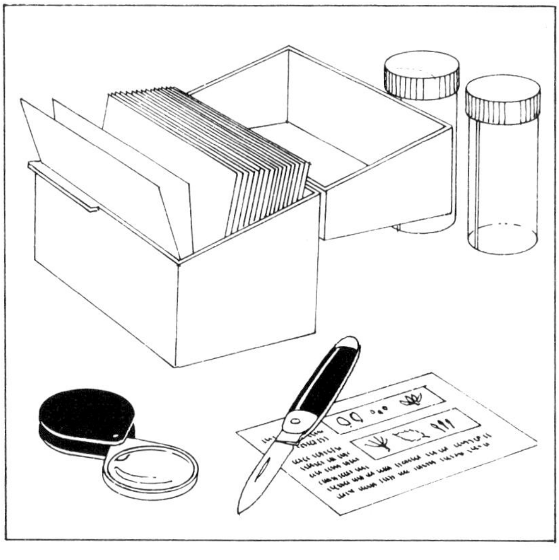

Some items of equipment

The year of the plant

There is a basic life cycle for most of the flowering plants that grow in this country. The seed germinates and grows to a mature plant capable of sexual reproduction; male and female sex cells are formed, and the male pollen is then carried to the female egg so that fertilization can occur. The formation of a new generation of seeds completes the cycle. Within that simple pattern, plants have developed a bewildering variety of methods by which they can reproduce.

Plants which complete their life cycle within one year are called annuals. They never have woody stems, and are often shallow rooted like Annual Mercury *(Mercurialis annua)*, a persistent weed of disturbed land and gardens in southern England. Annual plants are usually prolific producers of seeds, so that embryo plants can survive the cold and damp of winter, protected by a tough seed coat.

Plants with a life cycle extending from one year to the next are called biennials. They need to modify their structure to remain alive during winter, because those parts of the plant remaining above ground will be subjected to rain, frost, and snow. Many biennials form a rosette of leaves during their first year of growth. By their round shape and low profile, rosettes are particularly well adapted to withstand the rigours of winter as well as damage by grazing animals. Both Spear Thistle *(Cirsium vulgare)* and Great Mullein *(Verbascum thapsus)* develop impressive rosettes of leaves which serve to store food and nutrients, and from which the tall mature plant grows and flowers in the second year.

Where the life cycle of a plant is longer than two years it is called a perennial. Perennials tend to be stouter than annuals and, because of their persistent nature, they are able to flower each year if conditions are suitable. While some remain alive above ground during winter, others die back to a protected underground system, such as a bulb, corm, tuber or rhizome.

Bulbs, corms and rhizomes

These structures contain the food reserves which will enable the plant to begin again its growth when the winter ends. Bulbs are formed below ground by layers of swollen, fleshy scale leaves or leaf bases, which enclose the growth bud. A cross-section of a typical bulb, such as that of Grape Hyacinth *(Muscaria atlanticum)*, clearly shows the structure.

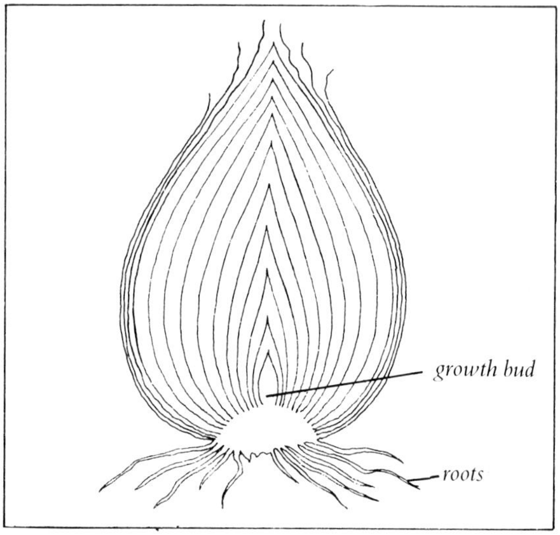

Bulb of Grape Hyacinth

Corms and tubers are swollen underground stems. A section through the corm of a crocus shows the old corm made during the previous year, which is responsible for the flower and leaf bud, with the new corm forming from the swollen stem above it.

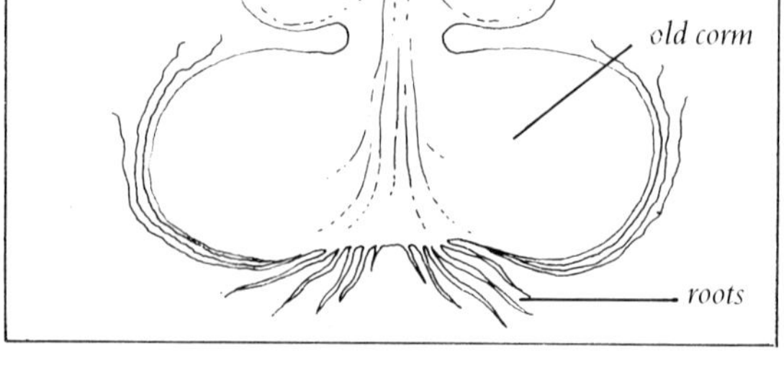

Crocus corm

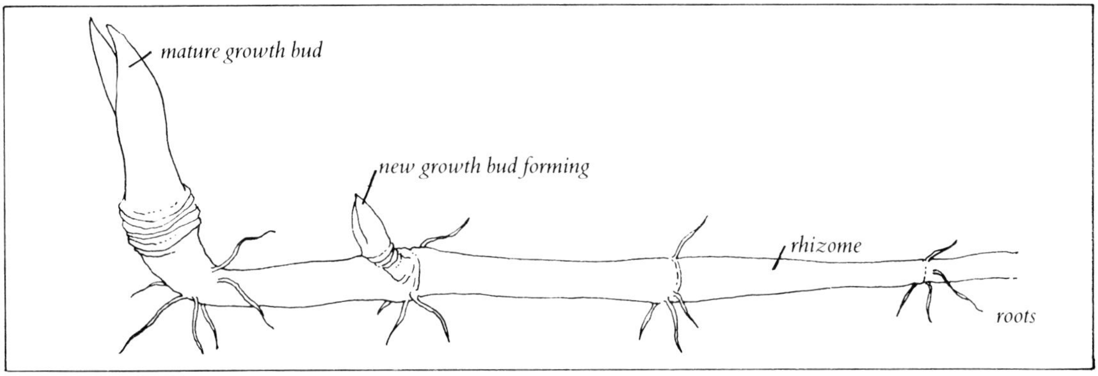

Lily-of-the-valley rhizome

Rhizomes are horizontal underground stems, and frequently are swollen and contain food reserves. The rhizomes of Lily-of-the-valley *(Convallaria majalis)* bear clusters of roots at the nodes, jointed swellings, with growth buds which produce leaves and flowers forming at the tip of the rhizome, and also at some of the intermediate nodes.

Flower structure

The flower is the reproductive apparatus of the plant. The whole structure is borne on the receptacle at the apex of the flower stem. At the base of the flower are sepals, often leaf-like in structure, the whole whorl of sepals comprising the calyx. Sometimes they are fused together to form a calyx tube with the sepals represented by calyx lobes.

Within the calyx lies the whorl of petals, which collectively form the corolla; they are usually brightly coloured and showy. Next come the male reproductive organs, the stamens, each consisting of a thread-like filament with the pollen-bearing anther at the tip. Finally, at the centre of the flower, lies the female organ, the ovary, containing one or more ovules. The ovary is joined by a column of united filaments, the style, to the pollen receptor at its tip, the stigma; the whole female structure is called the gynoecium. Flowers such as buttercups, which contain both male and female sex organs, are termed hermaphrodite but, in other species, the male and female organs may exist in varying degrees of separation. In the bur-reeds, reedmace, and duckweeds, the individual flowers are unisexual although flowers of both sexes bloom on the same plant; such a plant is called monoecious. In Dog's Mercury *(Mercurialis perennis)* male and female flowers are borne on different plants; such plants are termed dioecious.

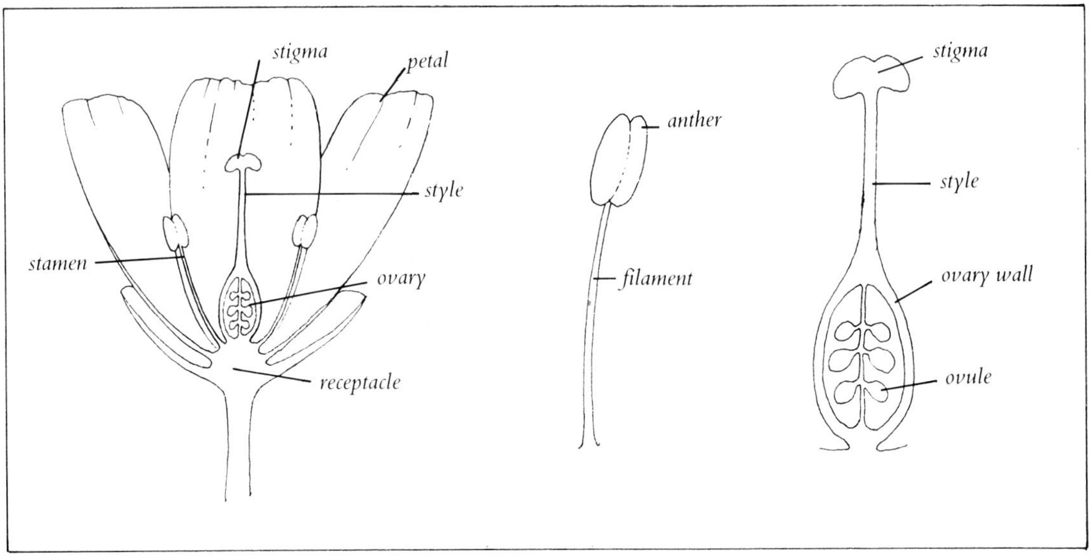

Section of a flower *Stamen* *Gynoecium*

Pollination

The male cells, pollen, are shed from the stamens. When the pollen grain reaches the stigma it germinates and a process, called the pollen tube, grows down inside the style to reach the egg in the ovule where fertilization occurs. In a few species, the flowers are capable of reproducing by seed which is not formed as the result of sexual fusion; such a process is called apomixis.

Photographs taken using a scanning electronmicroscope show that the microscopic structure of pollen grains is incredibly complex and beautiful. The pollen of each species is unique and can be recognized by its structure. This feature is used in the science of palaeobotany. Pollen grains are well preserved in peat bogs so that, if we study the pollen in the deep layers which were laid down thousands of years ago, we can build up a picture of the plants which grew there at that time, and hence the type of climate which prevailed.

Where pollen reaches the stigma within the same flower, the process is called self-pollination. Sometimes this can occur within the bud or unopened flower, when it is referred to as cleistogamy.

In many species the male and female organs mature at different times, or else the flowers are so constructed that it is impossible for them to come into contact without the action of some outside agency. Where the pollen is very fine and dust-like, cross-pollination between different individual flowers can be caused by the wind or, in aquatic plants, pollen may be carried mechanically by water from one plant to another. Many plants rely upon the activity of insects to carry the pollen from the anthers of one flower to the stigma of another. The scent, colour, and shape of the flower are designed in these cases to attract the insect pollinator, and flowers may produce nectar or other secretions upon which the insect will feed. The colour which the insect sees is not necessarily the same as that appreciated by the human eye; it is at the long wavelength, ultraviolet end of the spectrum. If flowers are photographed under ultraviolet light, bold lines and patterns become visible, and these guide the insect to the reproductive organs of the flower.

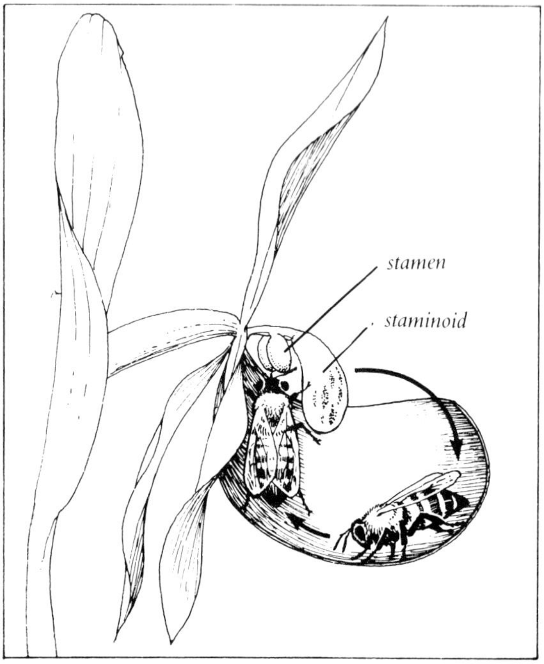

Pollination of Lady's-slipper

Flowers are often designed in such an manner that the visiting insect inevitably carries pollen from stamen to stigma. In this respect the various devices developed by our native orchids are some of the most complex and bizarre. The Lady's-slipper *(Cypripedium calceolus)* secretes nectar inside the yellow, pouch-shaped lip, which attracts small bees of the genus *Andrena*. The insect lands on the flat

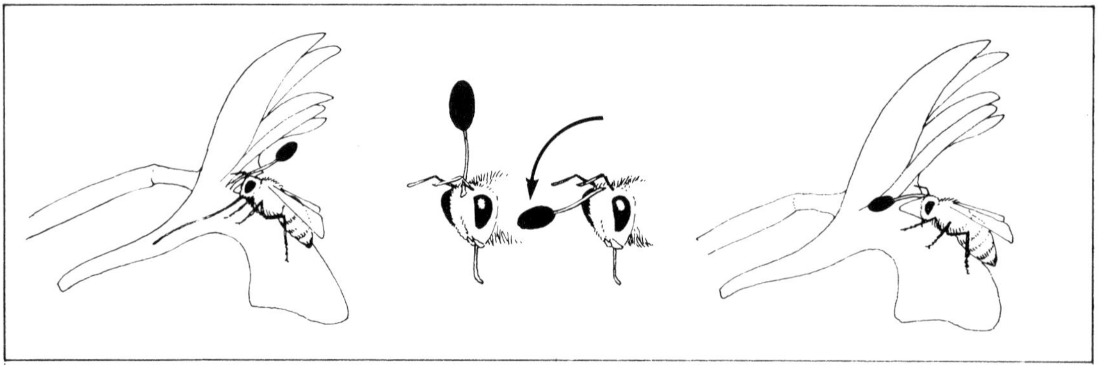

Pollination of Early-purple Orchid

surface of the staminode, and enters the pouch with ease, but is prevented from escaping by the curved, slippery wall. Near the base of the pouch, below the column, conveniently placed stiff hairs assist the bee to climb out. As it squeezes through the narrow exit, it brushes against the stamen and is dusted with pollen. When it visits the next flower, the pollen is brushed off on to the stigma which projects down into the slipper.

The two pollen masses or pollinia of the Early-purple Orchid *(Orchis mascula)* are borne on stalks which have a sticky disc, the viscidium, at the base. When the visiting insect attempts to remove nectar from the flower, the pollinia are glued firmly on to its head. Shortly after they are removed, the pollinia pivot through 90 degrees from the vertical to point forwards over the insect's head. In this position they strike the stigma of the next flower visited, and the time taken for this movement to be completed makes it likely that a different plant will receive the pollen, so that cross-pollination is ensured.

Pollination of the Fly Orchid *(Ophrys insectifera)*, whose flowers are convincingly insect-like, can occur by a process called pseudocopulation. The flowers attract male wasps of the species *Gorytes mystaceus*, which attempt to mate with the flowers, and carry away pollinia on their bodies to other flowers. This behaviour ceases when the female wasps, which emerge after the males, are plentiful.

Seed dispersal

Once fertilization has occurred, it is advantageous for the plant to spread its seed as widely as possible, so that it may have the best chance to germinate successfully. The mechanisms by which plants broadcast their seeds are fascinating and diverse.

Where the seed is very fine and dust-like, it can be dispersed by the wind, as is the case with many of the orchids. Wind dispersal is the method used by most members of the daisy family (Compositae), the dry seeds or achenes of flowers such as dandelions and thistles having a parachute-like pappus of fine hairs so that the whole structure drifts easily on the wind. Wind dispersal is also used by the willowherbs which release clouds of airborne seeds, each equipped with a fine plume of hairs, as soon as the ripe seed capsules are disturbed. Wind dispersal of a different sort is used by poppies. The ripe seed capsule has a series of openings or pores around the upper rim below the lid formed by the rayed stigma lobes. The hard, rounded seeds are scattered out through the pores in the manner of a censer, as the plant is shaken by the wind.

The burdocks and agrimonies have developed hooks and spines on their fruits, which catch in the fur of passing animals, or on our clothes, and are carried away and dispersed. The seeds of many water plants are very light and some, like those of White Water-lily *(Nymphaea alba)*, are equipped with hollow structures like floats, and are water dispersed.

Fruit of Dandelion

Fruit of Dwarf Thistle

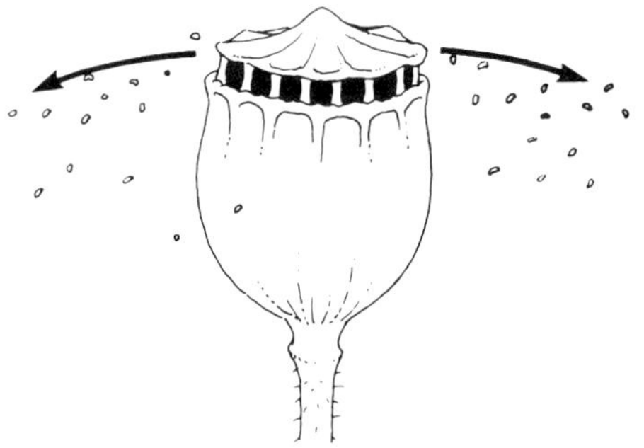

Seed capsule of Common Poppy

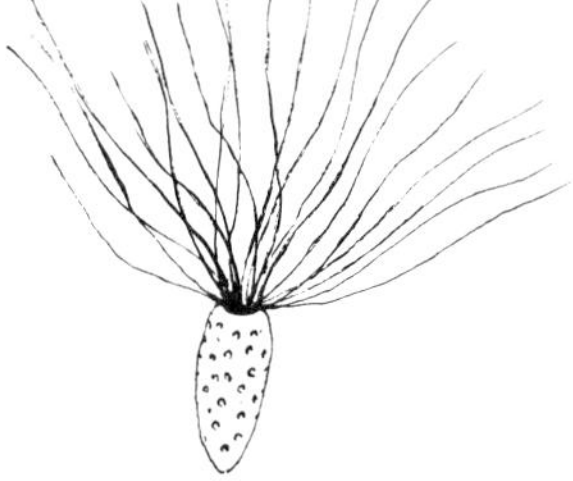

Fruit of Great Willowherb

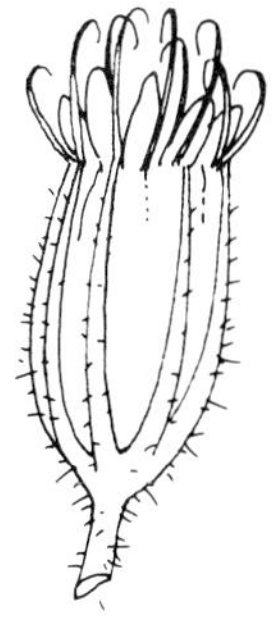

Fruit of Agrimony

Vegetative multiplication

Not all plant multiplication is accomplished by seed production for many plants spread and increase by the process of vegetative multiplication. Plants which produce rhizomes, such as Lily-of-the-valley, spread widely through the soil, with leaf and flower buds at intervals. These will then produce their own roots, and separation from the parent plant will give rise to satellites. Flowers which have bulbs or corms can form small secondary bulbs, which eventually detach and grow into separate individuals.

Drooping Saxifrage *(Saxifraga cernua)* rarely flowers, but forms bulbils in the leaf axils, and these fall to the ground, producing new plants. The Bog Orchid *(Hammarbya paludosa)* forms tiny bulbils in a fringe along the edges of the leaves. These break off and are dispersed by the surface water of the bogs in which the plant grows and give rise to new plants. The flowers of this species also set plentiful seed so the Bog Orchid uses two different methods of propagation.

Environmental factors

Each plant species thrives and flowers in a particular combination of temperature, light, and humidity. If you know what these conditions are, you will understand better why a plant should flower in a particular place at a particular time. No cellular activity can occur below the freezing point of water at 0 °C (32 °F). If you trace the lines of equal mean temperature during January and February in Britain, the only areas much above 4·5 °C (40 °F) are Cornwall, Devon, and the coast of Pembrokeshire, the west side of Anglesey, and the coast of Ireland from Dublin round the south coast to Sligo in the west. Only in these areas are we likely to find much plant growth in winter, and only there will it be worthwhile searching for flowers early in the year.

It is helpful to remember as a rough guide that every 300 metres' rise in altitude represents a drop in temperature of approximately 1 °C. Thus, the summit of Ben Nevis at 1343 metres will only remain consistently above freezing for the summer months, so that there is no point in looking for wild flowers at such an altitude except during that comparatively warm time. In fact, alpine plants growing under such frigid conditions may never produce seed, but rely upon vegetative means of reproduction to maintain and increase themselves. These plants are likely to be perennial because they are unable to complete their life cycles successfully in the brief period offered by the summer of a single season.

Humidity and available water are the next most important environmental factors. Bog plants, which need conditions of virtual saturation, are most likely to be found on the west side of England, Wales, and Scotland, and on the west side of Ireland, where the annual rainfall exceeds 150 centimetres. Similarly, those plants which flower in moist dune slacks, or among the seasonal freshwater turloughs of western Ireland, will not be at their best in the hot, dry months of the year.

Many of our carpeting woodland plants, such as Wood Anemone *(Anemone nemorosa)* and Bluebell, flower in that part of the year between the end of winter and the time when the trees overhead are in full leaf. Only in that interval will both warmth and light be sufficient.

The complexity of 'the year of the plant' will now be apparent. It may not even be completed within one season but will depend upon temperature, altitude, moisture, and sunlight. Plants can remain dormant as seeds for many years, or in a vegetative, non-flowering state, until conditions become suitable once more for flowering. This is particularly noticeable if a woodland is cleared or coppiced, when a host of plants suddenly make their appearance, to disappear slowly over succeeding years as the tree cover grows up again and the light is excluded.

In very cold or in very dry years, plants may remain dormant or fail to flower while, in exceptionally warm and moist years, plants may flower much earlier than anticipated, and continue to flower long after they would normally have finished. Even in a normal year the flowering period of the same species will vary by a month or more from the south of England to the north of Scotland, or even from the bottom to the top of the same mountain.

Where wild flowers grow

Influences on plant habitat

A line drawn from the River Tees to the River Exe divides Britain neatly into two; to the north and west lies the bulk of the high lands and mountains, whose rocks are the oldest in the country while to the south and east is lowland, with fertile plains overlying more recent sedimentary rocks such as limestone and chalk.

The base content of the rock, that is, the lime, potash, and magnesia within it, determines the richness of the flora which will grow upon it. If the rock is soft, then it erodes easily, releasing nutrients for the plants and, at the same time, forming gullies and sheltered sites for growth. The soft, basic mica schists of Ben Lawers are of this type, and account for the richness of its wild flowers. Hard rocks may contain all the necessary elements for plant growth, but these elements are not readily available, nor can the plants obtain a roothold.

The second important factor is temperature as we have already seen. Rainfall, too, has a marked effect upon habitat. Most of the rain-bearing winds come from the west, and the hill masses of the west cause much of the rain to fall there. High rainfall and relatively low temperatures discourage woodland growth, and lead to the formation of moorland and blanket bog, so that these types of habitat will predominate in the west. The light soils of parts of East Anglia contain little humus, and are easily leached of their basic elements, so that heaths occur mainly in the south and east, where the rainfall is lower. Man and his grazing animals have, over the centuries, shaped much of the country as we now know it. Burning and clearing have led to the disappearance of virtually all the primeval forest, while the presence of sheep has prevented the natural regeneration of woodland. Grazing animals created the fine pastures of the southern downland, and the machair grazings of the Hebrides.

Any habitat is a living community with plants and animals closely integrated; if this balance is upset the picture will be changed profoundly. In the natural course of events, things are constantly changing: trees grow to maturity, their fall creating clearings where smaller plants can flourish; streams and rivers erode the land; rocks shatter and fall; and tidal movements constantly reshape the shoreline of sand, mud, or saltmarsh. Man has an even more dramatic effect upon the land, draining, flooding, ploughing, and planting. For ease of description, the various types of habitat have been grouped systematically, but you should remember that the true situation is much more complex. It is the infinite variety within this framework which renders each site unique, and makes the study of plant habitat so fascinating.

Woodland

Woodlands seldom consist of a single species of tree, but a mixture of many, with an understorey of lesser shrubs, and a ground flora.

OAKWOOD

In the south, east, and central parts of Britain the oakwoods are mainly of Pedunculate Oak *(Quercus robur)*. and the heavy loam or clay tends to make the ground damp. Owing to the thick leaf canopy in summer, the flowers of oakwood are mainly perennial, and flower early while there is still enough light. Snowdrops, Ramsons, and Water Forget-me-not grow in damp areas, Marsh Thistle and Rosebay Willowherb in the clearings, while coppicing encourages Wood Anemones, Bluebells, and Primroses.

In the west and north of Britain, and on the sandy ground of Kent, Surrey, and Hertfordshire, the oakwood is formed by Sessile Oak, *(Q. petraea)*. There are few spring flowers apart from Bluebells, and grazing means that the trees are well scattered, with grass and bracken between them. Rowan and stunted Holly trees are common, with Heather and Bilberry on ridges and broken ground. Harebells, Common Cow-wheat, Tormentil, and Foxgloves are typical flowers of dry oakwood.

BIRCHWOOD

Downy Birch *(Betula pubescens)* is the common birch of the north and west because it is more tolerant of wet and cold, while Silver Birch *(B. pendula)* flourishes on drier heathland soils of the south and east. Lowland birchwood is fairly open and grassy, with Bluebells, Common Dog-violets, Primroses, and a scattering of Heath and Common Spotted-orchids between the trees. Where the wood is on

Scottish open pinewood

heathland, there is likely to be plenty of Bracken and Heather, with flowers such as Heath Bedstraw and Tormentil. In the Highlands, birchwood extends up to 700 metres, with a rich growth of Bilberry, Wood-sorrel, and mosses and, on occasion, the charming, white, star-like flowers of Chickweed Wintergreen.

PINEWOOD

The flower finder is well advised to shun conifer plantations, where close planting results in dense shade and few flowers. Only the Scots Pine *(Pinus sylvestris)* and Juniper are native conifers, and few areas of the ancient Caledonian pine forest remain in Scotland, except for remnants in Rothiemurchus, Abernethy and Glenmore Forests, and other scattered sites in the Spey Valley, the Black Wood of Rannoch, and a few places in north-west Scotland. The ground between the trees is fairly open, with a scattering of Juniper, Rowan, and Birch, and plenty of Heather, Bell Heather, and Bilberry. Beneath the Pine trees the fallen needles and moss form a moist, friable carpet, in which grow Creeping Lady's-tresses, Coralroot Orchid, and Lesser Twayblade with Chickweed Wintergreen, several of the true wintergreens, and the diminutive pink Twin-flower.

BEECHWOOD

Beech *(Fagus sylvatica)* grows mainly on the dry chalky soils of south-east England, on slopes with a southern aspect such as the escarpment of the North Downs in Kent and Surrey, the downs of Sussex, Hampshire, Dorset, Wiltshire, and the Cotswold limestone, especially near Cheltenham. Mature Beech trees are massive and cast a heavy shade; the leaves remain on the tree for a long time, so that undergrowth is sparse and there are few undershrubs. Beech leaves rot down to make a fine humus, in which grows the curious little saprophyte Yellow Bird's-nest.

Beechwoods are famous for the wealth of orchids they so often contain. White Helleborine is the most characteristic orchid of mature beechwood, with Narrow-leaved and the rare Red Helleborine in the spring, Fly, Lady, and Bird's-nest Orchids in summer, and Pendulous-flowered and Slender-lipped Helleborines in the autumn. The very rare Ghost Orchid grows in the depths of beechwoods, where the darkness has little effect, because it is entirely saprophytic.

ASHWOOD

Ash *(Fraxinus excelsior)* grows best on calcareous soils. In south and central England, where it grows with Beech, the Beech will ultimately overtop it but, on the Carboniferous limestone of the Derbyshire Dales, West Yorkshire, Ross, and parts of Mendip in Somerset, Beech will not grow, and Ash forms fine open woodland with Bird Cherry, Aspen, Whitebeam, and Wayfaring-tree. The ground flora is fairly rich and includes Moschatel, Ramsons, Wood Anemone, Lesser Celandine, and Lily-of-the-valley. Rassel Wood at Kishorn in Ross-shire is an extraordinary mature ashwood growing out of cracks between massive, moss-covered blocks of limestone dotted with clumps of Primroses and Early-purple Orchids. Similar woods exist at Colt Park wood in west Yorkshire and Tokavaig on Skye.

ALDER CARR

Alder carr forms from reed swamps in wet valley bottoms where the water is slow flowing. Seedlings of Alder *(Alnus glutinosa)* and willow root in the tussocks, which make the ground so difficult to negotiate, and soon, other trees such as Buckthorn and Alder Buckthorn help to form a dense scrub. Clumps of Nettles and Common Comfrey, with tangles of brambles, make life even more difficult for the botanist, but the reward is to find flowers such as

Southern Marsh-orchid which grows to a great size in this habitat, and the massive, broad-lipped form of the Fragrant Orchid.

HAZEL COPPICE

Coppiced Hazel *(Corylus avellana)* is commonest in the south of England, particularly in Kent and Dorset, the delicate tracery of leaves permitting a moderate amount of sunlight to filter to ground level. This results in a rich flora with masses of Bluebells, Ramsons, Red Campion, Moschatel, Toothwort (which is parasitic on the roots of Hazel), and orchids such as Early-purple and Common Spotted-orchid, Greater Butterfly-orchid and Fly Orchid. In Kent, Lady Orchid may also be found and, in Dorset, I have seen a Hazel coppice so full of Common Twayblade that little else could grow on the floor of the wood.

Hedgerows, ditches, and verges

HEDGEROWS

Most hedgerows are constructed with a bank and ditch, the hedge planted on top of the bank, and the ditch serving to drain the adjacent field. The main function of a hedge is to contain stock and, for this reason, the most popular hedge is a mixture of Hawthorn, Hazel, and Blackthorn, because they are all quick growing. But hedges may consist of a wide range of species including Privet, Holly, Berberis, Elder, Snowberry, Dogwood, and Guelder-rose, the latter two being common on chalk and limestone soils. Many trees are used in hedge construction; Wych Elm is common in East Anglia while Beech occurs on Exmoor. Sycamore and Field Maple are widely used in the south, with Rowan, Wild Cherry, Crab Apple, Pedunculate Oak, and Lime all finding a place as hedge plants, some growing to full tree stature in the hedge.

Traveller's-joy and Black and White Bryony are common climbers in hedges in the south on chalk soils, while roses and brambles of all sorts grow in hedges throughout Britain, together with Bittersweet, Field and Hedge Bindweed, and Honeysuckle. The three climbing vetches, Bush and Tufted Vetch and Hairy Tare, add a bold splash of mauve and blue to the summer hedges.

The banks below the hedges are fine places for seeking plants, offering a vast range of habitat. A warm, south-facing bank, for example, is the first place you will find violets and Primroses in the early spring. The dryness of a hedge bank is an important factor in determining which plants will grow there. Shepherd's-purse, Common Chickweed, Dandelions, Dog's Mercury, and Garlic Mustard all flourish on dry banks in spring. Damp banks are the places for Primroses and Cowslips, Common Dog-violet and the scented Sweet Violet, Lesser Celandine, Bluebells, Lords-and-ladies, and much else.

In summer, Greater Celandine, Dove's-foot Crane's-bill, Yarrow, Wood Sage, and the tall yellow Great Mullein favour the dry spots, while Red and White Campion, Cut-leaved Crane's-bill, Wood Avens and the tall, sombre Hemlock flower on banks which are moist. These are only a few of the many flowers you can find.

DITCHES

The ditches below hedgerows offer a variety of sheltered and shaded habitats, too. The depth to

Typical southern hedgerow

which the ditch is dug, and its wetness, have a marked effect on the plants to be found there. Most of the ditch plants are annuals which can quickly re-establish themselves after a ditch is cleared out, or deep rooted perennials, which can survive the clearing operation.

Early in spring, dry ditches are the places to look for Colt's-foot, followed later by Common Chickweed, Red and White Dead-nettles, Black Horehound, and Silverweed. Where it is a little damper, Nettles and Common Comfrey flower early, followed by Hogweed and Broad-leaved Willowherb. In high summer the ditches are full of thistles whose purple flowers attract many feeding butterflies. Marsh marigold is one of the first plants to flower in damp ditches, followed by Water-dropwort, Fool's Water-cress, the blue flowers of Water Forget-me-not and Brooklime, and, finally, exuberantly growing plants such as Meadowsweet, which swamp everything else.

VERGES

Road verges have their own distinctive flora although it is greatly modified by trimming and the use of herbicide sprays. In springtime Winter Heliotrope, Colt's-foot, and Dandelions flourish better on verges than anywhere else and, during the summer, verges are a prime site for clovers and related species, such as Red, White, Alsike, and Hare's-foot Clover, Lesser Trefoil and Black Medick, Restharrow, and the slender, red-flowered Grass Vetchling. Common Knapweed, Creeping and Spear Thistles are all widespread on verges and, in late summer and autumn, the dandelion-like flowers of Goat's-beard, hawkweeds and hawkbits. Road verges enjoy their fair share of rarities, such as Spiked Rampion in east Sussex and Sulphur Clover in East Anglia. The increased use of salt to de-ice roads has led to the spread inland of many salt-tolerant plants which are usually associated with the seaside.

On the field side of the hedgerows, where the ground has been ploughed and arable crops planted, a whole range of different plants can be found – remember to keep off the crops! Shepherd's-purse, Wall, Ivy-leaved, and Common Field-speedwell are common, with Field Penny-cress, Common Chickweed, and Corn Spurrey. On the edge of cornfields look for poppies, Field Pansy, Common Toadflax, and Venus's-looking-glass.

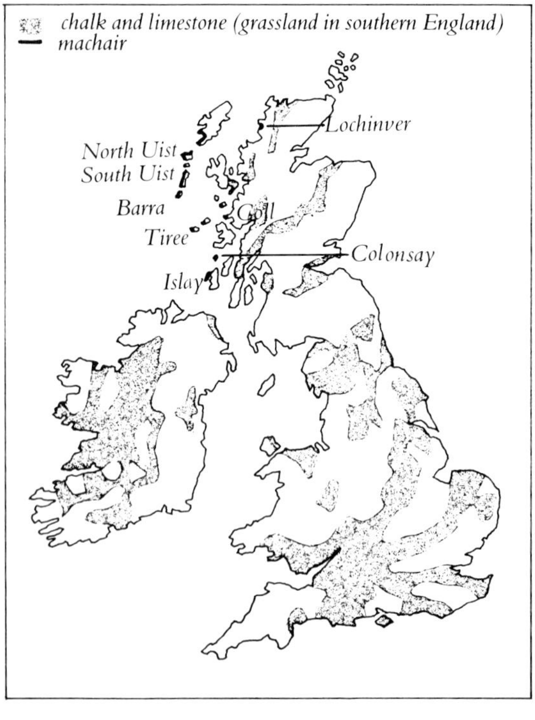

Distribution of calcareous soils in Britain

Grassland

Grassland is an entirely man-made habitat; its composition depends upon the nature of the underlying soil.

NEUTRAL GRASSLAND

Neutral grassland exists on the fertile clays and loams of central and south-east England, where the soil is neither alkaline nor acid. It has been formed where forest was cleared for cultivation and, if left untended, would revert to forest again, but it is grazed, or cut for hay and silage. It can be divided into permanent pasture, and leys which are ploughed and reseeded with grasses and clovers intended for short-term use over a period of several years. Flowers and weeds, such as buttercups, thistles, and plantains, establish themselves in these pastures by virtue of their unpalatability to grazing animals. In damp areas Cuckooflower, Creeping Buttercup, Marsh-marigold, and Meadowsweet all flourish and in a few unspoiled wet meadows in the Thames basin, Fritillary still flowers in abundance.

CHALK GRASSLAND

Chalk grassland is one of the most beautiful habitats that the botanist can visit. It occurs on the South Downs of Sussex, the North Downs of Kent and Surrey, parts of the Isle of Wight, the grassland of Wiltshire, Hampshire, and Dorset, and the downs of Oxfordshire and Berkshire. The soil is dry and well drained, and grazed by sheep and rabbits, so that many of the plants are prostrate in form, or grow with tight rosettes of leaves. It has a sweet scent from Thyme and Marjoram. The downland turf was formed by sheep but few areas of old downland remain undisturbed by the plough. The recent marked rise in the sheep population may well herald a return of downland grassland to something like its state before 1939.

It is useful to acquire an 'eye' for areas of chalk grassland where the sward is fine, with an almost bluish tinge from the grasses growing there, and areas which have never been ploughed because of the steepness of slope; there the flora will be rich. Even small areas above cultivated slopes can be remarkably rewarding, such as the sides of earthworks and barrows, and the floors of abandoned chalkpits. Cowslips, Early-purple Orchids, the rare Pasqueflower, and Early Spider-orchid are among the earliest flowers, followed by Chalk Milkwort, Kidney Vetch, Squinancywort, and Salad Burnet in full summer. From late May to July the downland orchids are at their finest, with Burnt Orchids, Fragrant Orchids, and Common Spotted-orchids, and then Man, Bee, and Pyramidal Orchids. Late summer is the time for Musk and Frog Orchids and, at that time, all the thistles and scabious are in bloom, with Greater Knapweed, Round-headed Rampion, and Yellow-wort. Finally, in autumn, Clustered Bellflower, Autumn Gentian, and Autumn Lady's-tresses bring the season to a close.

LIMESTONE GRASSLAND

The limestone grassland of the Cotswolds, Pennines, Lake District, and parts of north Wales has a very similar flora to the chalk grassland of the south. In spring, it has its own specialities such as Mountain Pansy and Bird's-eye Primrose, with Lesser Butterfly-orchid more frequent than it is in the south and, rarely in north Wales and Yorkshire, the diminutive Small-white Orchid.

UPLAND GRASSLAND

The grassland of the uplands of the north-west and north is dominated by the grasses Sheep's Fescue and the bents. It is usually somewhat acid, with extensive areas of Bracken, and is floristically dull. Creeping Buttercup, Heath Milkwort, Tormentil, Sheep's Sorrel, Heath Bedstraw, and Heath Speedwell are all widespread and common.

HILL MEADOWS

The hill meadows of northern England, Wales, and Scotland make up a very special habitat; they are usually small, permanent pastures surrounded by stone walls. The difficulty of access prevents the use of heavy machinery, little artificial fertilizer is used, and, since only a single hay crop is taken in late June or July, the meadow flowers have a chance to set seed. Consequently, the flora is rich and diverse. One small meadow near Dolgellau in Gwynedd held nine species of orchids along with many other delightful and colourful flowers, and a similar picture can be seen in Scotland, where hill meadows are an excellent habitat for the Small-white Orchid.

MACHAIR

Machair is a seminatural grassland growing on shell sand, to be found mainly on the Uists, Barra, Tiree and Coll in the Hebrides, Colonsay and Islay, and on the Scottish mainland at Ardnamurchan Point and Lochinver. There the winters are mild and the summers cool, with strong winds and a rainfall of 100 centimetres. The machair forms a green belt on

Hebridean machair

the western coast of the islands, dotted with the white cottages of the crofting townships, and lying between the Atlantic and the heather heath inland. It is a beautiful place, the result of centuries of carefully controlled grazing by cattle and sheep. In the past the crofters of each township would appoint a constable, whose job it was to ensure that no excessive grazing occurred. If the fine turf is disrupted and bare sand exposed, then the fierce winds rapidly create an area of blow-out, the sand ruining the grass for metres around. The flowers of the machair are its glory, and exist in a profusion which is remarkable. All the chalk downland flowers which would be familiar to a southerner are there, with Bloody Crane's-bill, Sea Stork's-bill, Sea and Buck's-horn Plantain, medicks and vetches in profusion, Fairy Flax, Wild Pansy, Seaside Pansy, sheets of Eyebright, and, in the damper hollows and drainage runnels, Yellow Iris and Grass-of-Parnassus. The brick-red, sand-dune form of Early Marsh-orchid and Broad-leaved Marsh-orchid are widespread in spring, with many Frog Orchids in late summer, but pride of place must go to the Hebridean Orchid, a dwarf form of Common Spotted-orchid peculiar to the machair, flowering in profusion throughout June and July.

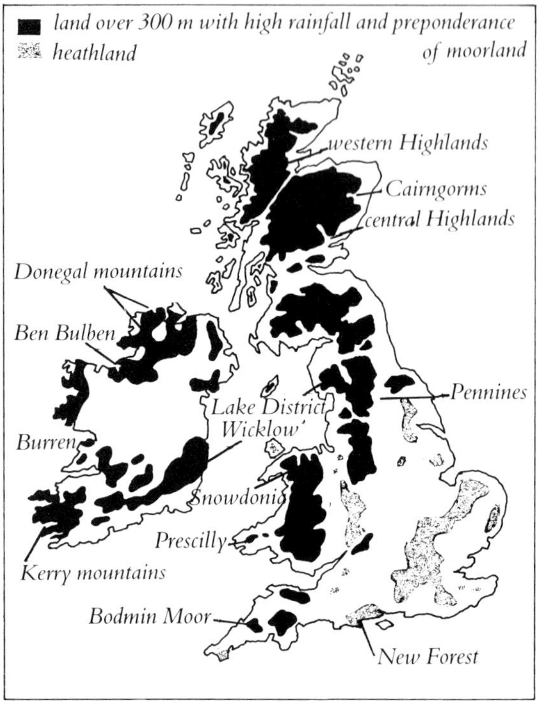

Highlands and heaths

Heath and moor

HEATH

Heath forms on light, sandy or gravel soils in areas of low rainfall. Extensive heathlands occur in lowland England on the Ashdown Forest in Sussex, the Surrey commons, the New Forest of Hampshire, and the heathlands of Dorset, east Devon, and the Lizard in Cornwall. Minerals are easily leached from the soil, which contains little humus, and the thin surface layer of peat quickly dries out in the sun. Heather and Bell Heather are dominant plants on the heathland, with extensive areas of Bracken. The rarer Dorset Heath grows on the heaths to the west and south of Poole Harbour, as well as in Devon and west Cornwall, and the Cornish Heath on the Lizard peninsula. Look for the curious parasite, Dodder, draping its pink, spider's-web strands over the gorse. Fairy Flax, Tormentil, Heath Milkwort with its slim, dark-blue flowers, and Bilberry are all common on heathland, while August is the time to look for the rare Heath Lobelia, which is restricted to a few southern heaths.

In damp, boggy areas in spring Marsh Violet and Early Marsh-orchid can occasionally be found, the latter usually in the mauve colour form, var *pulchella;* the very rare yellow form var *ochroleuca* still flourishes in one heathland site in the south. The pale-pink spikes of Heath Spotted-orchids are common on southern heaths in summer, and the damp places usually contain Common Butterwort, Round-leaved and the rarer Great Sundew. Marsh Gentians flower in early autumn in moist, grassy areas, and not in the marshy spots which their name suggests.

The Breckland is an area of about 650 square kilometres of heathland in east Suffolk and Norfolk. The soil was formed from sandy glacial material from which the limestone and fine clay particles were washed out, and it is named after the 'brecks' or 'brakes' which are patches of land cleared years ago of trees and heath vegetation for temporary cultivation. Few extensive areas of breckland now remain, much of the land being planted with pine trees, ploughed, or taken over for military use. The Breckland is famous for the rare Breckland and Spring Speedwells in the spring, Spanish Catchfly in summer, and Spiked Speedwell in autumn, and a wealth of other plants now sadly much reduced in number, and hard to find.

MOOR

Moorland forms in places where the underlying rock is acid and the rainfall is high. Therefore, it is commoner in the north and west of Britain, the areas

Typical Scottish or Yorkshire moorland

corresponding in the main to the high land over 300 metres. Extensive expanses of moorland occur on the low, eastern flanks of the Scottish Highlands and the Pennines, and moorland predominates in much of north-east Yorkshire and Galloway. Although it is splendid walking country, moorland is often poor in flowers. Moors and heaths are examples of an extreme type of habitat to which few plants are able to adapt and, for this reason, large areas become dominated by one type leading to a uniformity in the vegetation.

The activity of sheep and the rotational burning of heather moors destroy shallow-rooted plants and encourage Heather and Gorse. The shallow, sandy soil does not drain because of the impervious rock underneath, peat accumulates, and the soil is deficient in nitrogen. Gorse is able to thrive through the presence of nitrogen-fixing bacteria in nodules on the roots. Heather, Bell Heather, Cross-leaved Heath, and Cowberry all have symbiotic fungi living in their roots which make nitrogen available to them, and plants such as the butterworts and sundews obtain their nitrogen from the bodies of insects trapped in their leaves.

Although the heaths, heathers, and Gorse dominate much of the land, there are still many flowers to be found, Harebell, Heath Bedstraw, Tormentil, Heath Spotted-orchids, Crowberry, and Cowberry are all common. On the moors of the eastern Highlands, the little gorse-like Petty Whin and Intermediate Wintergreen are widespread, with Lesser Twayblade in damp moss under the heather among the rocks while, in the north-west of Scotland on damper moorland, grow Dwarf Cornel and Cloudberry, with its delicious orange berries.

Mountains

The mountain flowers of Britain are as varied and superb as the places where they grow. The diversity of habitat is astonishing, and it is possible here only to touch the surface of a fascinating subject. The splendid book, *Mountain Flowers*, by John Raven and Max Walters is recommended reading for all who walk and climb in the hills, and is alive throughout with the special joy of finding mountain flowers.

Rainfall, temperature, and rock all have a profound influence on the mountain flora. On the high land, water drains away rapidly because there is little humus to hold it, so that areas of high rainfall will be richer in plants. Temperature is vitally important, as we have seen. The amount of basic material in the rock, and the softness of the rock which renders it available to the plants, have a profound influence. Most upland soil is acid, covered in peat or bog, so that few species are abundant over large areas. The best places for flowers will be those areas of basic rock and cliff inaccessible to sheep, on slopes steep enough to avoid a blanket covering of moorgrass or bog, but not necessarily on the mountain tops.

Lime-rich mountains occur in England in Teesdale, on Ingleborough and Pen-y-ghent, in Wales on Stanner Rocks, Great Orme's Head, parts of Cader Idris and Snowdon, and in Ireland in the Kerry Mountains, Ben Buishgen, and the Ben Bulben Range north and west of the town of Sligo. Pride of place must, however, go to the mountains of the west Highlands, the western isles, Skye and Rhum. There is a magnificent stretch of country from the Breadalbanes in Perth to Glen Doll in Angus, including Ben Lawers with its fabulous flora, Lochnagar, Glen Clova, the limestone of the Knockan Rocks, Inchnadamph and Ben More, and the Sutherland limestone of Betty Hill and Durness. On many of the rather flat summits of the Cairngorms, the acid grassland gives way to an area of Crowberry and Bilberry, and finally to a bare open summit with sparse plant growth. This zonation of plants with altitude is not always clear cut and, in the far north of Sutherland, the mountain heath type of vegetation, normally found over 1000 metres, descends to sea-level. Three broad zones can be recognized. The lowest zone consists of dwarf Juniper, Heather, Bearberry, and much Deer-grass. The middle zone, which is at an altitude of 400 to 900 metres in the north Highlands and 800 to 1200 metres in Cairngorm, consists of a flora dominated by Heather, with Black Bearberry and Trailing Azalea, grading into Bilberry, Crowberry, and Alpine Lady's-mantle at the upper limit. The summit vegetation at 700 to 1300 metres consists of Wavy Hair-grass with club mosses, and alpine plants such as Dwarf Willow, Alpine Lady's-mantle, Moss Campion, Cyphel, and Alpine Bistort. On the base-rich mountain summits, Starry, Alpine, and Purple Saxifrages flower, with Alpine Forget-me-not and Rock Speedwell.

At high altitude, sharp alteration between day and night temperatures coupled with rapid drainage of moisture and high winds, lead to plants whose form is dwarf, with tight rosettes of tough, resistant leaves. Many of the species are flowers left behind by the retreating ice of the last Ice Age, and Fraser Darling has aptly described the Highland summits as '...an immense archipelago of biological islands holding relict communities...of the last postglacial epoch.'

It is often worthwhile to concentrate your efforts, not on the summit, but on the broken ground below, especially where sheep do not graze. Each area will tend to have its own special flowers. Sheltered cliff ledges on the mica schist mountains have a rich flora which includes Roseroot, Globeflower, Mountain Sorrel, Wood Crane's-bill, Alpine Meadow-rue, and Alpine Saw-wort. Sheltering under boulders and in crevices are Mountain Sorrel and Wood-sorrel, and Opposite-leaved Golden-saxifrage. On open ridges, exposure leads to the presence of plants with rosette or cushion form, including Cyphel, Starry Saxifrage, Alpine Lady's-mantle, Trailing Azalea, Dwarf Cudweed, and a dwarf form of Goldenrod; and, on just such a bleak ridge in Wester Ross, the dwarf Norwegian Mugwort was found in 1952, new to Britain.

On the edges of mountain streams, Blinks, Starry and Yellow Saxifrages, Lousewort, and Alpine Willowherb grow, while mountain lochs are the place to look for Water Lobelia and White Water-lily. Rockfalls are often covered with a mass of Bilberry, Heather and mosses. Look in the moss under the Heather for the diminutive Lesser Twayblade and, on the steeper pitches of base-rich rocks, for Mountain Avens, Purple and Mossy Saxifrages, and Roseroot. Never neglect the foot of a cliff face – it is often rich in plants.

LIMESTONE PAVEMENTS AND CLIFFS

Carboniferous limestone occurs in northern and western England, Wales, central Scotland, and western Ireland. Other limestones are found from north Yorkshire to Dorset, while the most recent calcareous rocks are the soft chalks of southern and eastern England. However small the limestone area may be, it is almost certain to be worth visiting. Limestone pavement is formed of horizontal sheets of limestone, deeply dissected by water action into massive flat blocks, with deep crevices, or grykes between them. It is in the sheltered grykes that humus accumulates and flowers grow. On the limestone of the Gower in south Wales, Great Ormes Head, and the Severn valley in east Montgomeryshire, Goldilocks Aster and Spiked Veronica grow, with Dark-red Helleborine on Great Orme's Head. The limestone of Yorkshire, Teesdale, and the Lake District also possesses Dark-red Helleborine, with Mossy and Purple Saxifrages, Mountain Avens, Alpine Cinquefoil, and the poisonous Baneberry in the grykes of the limestone pavement.

Limestone pavement

Southern stream side

In Teesdale, Hoary Rockrose, Spring Gentian, and Shrubby Cinquefoil are specialities, a combination repeated on the superb and extensive limestone pavement and cliffs of the Burren in County Clare and Galway in western Ireland. There, Mountain Avens and Hoary Rockrose are abundant and, in addition to Dark-red Helleborine, the Burren has its own specialities of Dense-flowered Orchid, many unusual marsh-orchids, Large-flowered Butterwort, and masses of Bloody Crane's-bill.

The limestone cliffs and rocks of Kishorn, Inchnadamph, and Durness in north-west Scotland are superb sites for Dark-red Helleborine, with sheets of Mountain Avens and saxifrages of many species.

Freshwater habitats

Plants of freshwater habitats can be broadly separated into those which grow entirely in water and those which grow along the water's edge. Aquatic plants such as the water-crowfoots often have two different types of leaf: submerged leaves, which are finely divided so as to offer minimal resistance to water flow, and surface leaves which are large to allow a maximum area for metabolism.

Few plants are able to grow in fast-flowing water, where silt does not settle, and the rock bottom affords no hold for their roots. Slow-moving streams, canals, ditches, ponds, and lakes are all rich areas for plant hunting. Out in the deeper water, growing from roots deep in the mud, are White and Yellow Water-lilies, both with broad, flat, floating leaves. Plants of less deep water include Water-violet, Frogbit, Bogbean, the many pondweeds, Amphibious Bistort, and Water Lobelia. The shallower water of the margins is the place to find Water-cress, Common Water-crowfoot, and, floating free on the sheltered surface, Water Fern and the tiny green duckweeds.

The water's edge supports a vast assembly of plants, either rooting in the water or in the wet mud beside it. The list of species is large, ranging from waterside plants such as Fool's Water-cress, Water Dock with its huge pointed leaves, Water Mint, Water-plantain, and Flowering-rush, to plants of the damp land such as Marsh-marigold, Lesser Spearwort, Cuckoo-flower, Indian Balsam, Meadowsweet, Great Willowherb, Water Forget-me-not, Monkeyflower, and Yellow Iris. June, July, and August are the best times to visit waterside areas to see wild flowers.

Streamside flowers of hill and mountain districts will only flourish where sheep cannot reach them and, beneath the Birch, Rowan, and Oak grow shade-loving plants more commonly associated with damp woodlands, such as Ramsons, Wood-sorrel, Bluebells, and Primroses. The flowers of lakes and lochs are governed by the depth of the water, and whether it is lying over a peaty or an alkaline soil. The dubh lochans of Sutherland and Argyll lie in areas of gneiss, and contain few plants of interest except White Water-lily and Bogbean in the deep water, various pondweeds, Lesser Spearwort, and Water Lobelia in the shallows, and Shoreweed on the edge. By contrast, the flora of the shallow limestone lochs of Lismore, between Oban and Mull, is rich, with Common Water-crowfoot, Lesser Spearwort, Marsh-marigold, White and Yellow Water-lilies, Bog Stitchwort, Meadowsweet, Purple-loosestrife, Marsh Pennywort, Marsh Ragwort, Bogbean, Greater Bladderwort, many pondweeds, and Yellow Iris.

MARSH

Marsh consists of slow-moving or static water of neutral *p*H, overlying soil containing a little peat, and much rotting vegetation. There is a profuse growth of reeds and rushes, with moss between them, as well as Marsh-marigold flowering in the early spring and Water Mint in summer. It is for their orchids that marshes are famous; Early Marsh-orchid is the first to flower, the summer species depending upon the part of Britain involved, finishing with Marsh Helleborine in early autumn.

FEN

Fen is formed where peat is covered by neutral or alkaline water derived from chalky rocks. Fens are found in East Anglia, in Suffolk, Norfolk, and north Essex; as well as in Anglesey, at the head of Esthwaite Water in the Lake District, and on the shores of Loch Neagh in Ireland. In the fen district of East Anglia, only a few fragments of the original extensive fens remain. Drainage was started by Cornelius Vermuyden, who constructed the Old and New Bedford Rivers in the seventeenth century, to drain the fens out into the Wash and, since that time, there has been an extensive drying out of the peat. An iron post driven level with the ground into the underlying clay in the mid-nineteenth century now stands exposed for 4 metres because of the shrinking of the peat.

The fenland habitat of the nature reserves at Holme, Woodwalton, and Wicken Fens has to be carefully conserved if it is not to deteriorate, and periodic reed cutting is essential to preserve the balance of the flora. In addition to such flowers as Water Violet, Greater Bladderwort, Fen Violet, and Milk Parsley (the food plant of the Swallowtail Butterfly), the fens contain some rare orchids including Fen Orchid, the dense-flowered form of Fragrant Orchid, Irish Marsh-orchid and, until recently, the rare yellow form of the Early Marsh-orchid. The Broads are shallow, lake-like extensions of the lower courses of the Ant, Bure, Yare, and other rivers, with margins of reed swamp and small areas of fen-type vegetation.

BOG

Bogs are formed when acid water overlies a peaty base, and are of three types.

Valley bogs occur in shallow depressions in wet heathland, and can be found even in the dry eastern part of England. They are the typical bogs of the New Forest and Surrey heaths, and contain a carpet of sphagnum moss, with Cross-leaved Heath, Bog Myrtle, Round-leaved and Great Sundew, Marsh Cinquefoil, butterworts, Bog Asphodel, and Bog Orchid, Heath Spotted-orchid, and Lesser Butterfly-orchid.

Raised bogs develop on top of valley bogs through the continued growth of sphagnum moss and the formation of peat. The whole bog is domed, more acid in the centre, and has fewer plant species as well as a surface broken by pools and hummocks. Bilberry and Heather grow in drier areas but otherwise the flora is similar to that of a valley bog.

Blanket bogs form in areas of high rainfall and humidity and have a covering of sphagnum except where the drainage is good. In the western Highlands they form on gentle slopes below 500 metres and are also found in Wales, the Pennines, Dartmoor, and Connemara in Ireland. In patches of open water, Bogbean, pondweeds, and bladderworts are plentiful, with butterworts and Bog Orchids in mossy areas, and Bog Asphodel with Bog Myrtle in the drier places. Heath Spotted-orchids and Lesser Butterfly-orchids are not uncommon, together with the occasional Northern Marsh-orchid. In the eastern Highlands the blanket bogs contain more Heather, Bilberry, Crowberry, and Cloudberry, and far less Bog Myrtle. They form above 300 metres especially in the Monadliadhs, Atholl, and Angus and, occasionally, one may find Cranberry and Small Cranberry. Dwarf Birch and Black Bearberry appear in blanket bogs above 800 metres in the Cairngorms and on Lochnagar and, in these high-level blanket bogs, Heather is replaced by Crowberry. All bogs are fragile habitats and great damage can be done by over-enthusiastic flower seekers squelching around without proper care for the plants. Walk delicately, if only to save yourself from unplanned immersion!

Seaside habitats

There are some similarities between the treeless environments of seashore and mountain top, and a number of flowers, including Thrift, Sea Campion, and Sea Plantain, are common to both. The most important factor influencing seashore plants is salt. An excess of salt interferes with the fluid balance inside the plant and seashore plants, despite their wet environment, have many characters of xerophytes, that is, plants of dry areas. Both groups have foliage modified to prevent water loss. Marram and Thrift have stiff, rolled leaves to reduce surface area; scurvygrasses have leaves with glossy, resistant surfaces, while Roseroot and the stonecrops have closely crowded, succulent leaves. The flat leaves of Common Orache and Sea Campion show little adaptation to a salt-laden environment.

Maritime habitats

SALTMARSH

In saltmarsh there is a sluggish outflow of fresh water and a twice-daily incoming tide of salt water. At high tide these balance, and there is virtually no movement, so that fine silt particles settle out, forming flats and banks of mud. At the lowest high-tide level, Glasswort grows with Sea Plantain. Higher up there is a zone with Annual Sea-blite, Sea Aster, Sea Milkwort, and Sea Arrowgrass. The highest zone, which is only covered by the spring tides, contains Thrift, Common Sea-lavender, Sea Plantain, sea-spurreys, and Marsh-mallow in the brackish ditches behind the saltmarsh.

SANDY SHORE

The shore above the high tide is the place to look for Sea Milkwort, Saltwort, Glasswort, Sea Spurge, the prickly blue Sea Holly, Sea Bindweed, and Sea Sandwort. Sea Couch-grass and Marram play a vital role, binding the shifting sand to form stable dunes.

DUNES

When a dune is fully covered by vegetation it is called a fixed dune, and the low-lying damp areas are called dune slacks. Sand derived from sea shells is rich in calcium, and calcareous dune slacks are fascinating places to search for flowers, which include Sea Stork's-bill, Cat's-ear, Spiny Rest-harrow, Carline Thistle, Squinancywort, Silverweed and Creeping Cinquefoil, Early Forget-me-not, Biting Stonecrop, Thyme, many different clovers and trefoils, and a delightful form of Heath Dog-violet with blue flowers and a yellow spur. Seaside Pansy grows in dunes in the north-west.

The orchids of dune slacks are superb, including Green-veined Orchid, a beautiful bright-red form of Early Marsh-orchid, many of the spotted and marsh orchids and their hybrids, Pyramidal Orchid, Marsh Helleborine, and a number of very rare plants such as the broad-leaved form of Fen Orchid in south Wales and north Devon, Coralroot Orchid in east Scotland, and Dune Helleborine in its classic localities on Anglesey and in Lancashire.

SHINGLE

Despite the bleak and forbidding appearance of a shingle beach in winter, and its apparent barrenness and lack of soil, it is a rewarding place in which to seek wild flowers, especially in the period from May to the end of July. Most of the flowers which grow there grow in other habitats, too, although Bittersweet, Blackthorn, and Broom all grow in curious, prostrate forms on the shingle.

Typical of shingle beaches in the south are the cabbage-like Sea Kale, the beautiful Sea Pea, and Yellow Horned-poppy, with orange-yellow flowers and curved seedpods from which it gets its name.

SEA CLIFFS

The flowers of sea cliffs will vary according to the type of rock forming the cliff, the distance above the splash zone of salt water, and whether the cliffs are in the north or south of Britain. Sea Campion, Sea Beet, and Thrift are common in all areas, as is Roseroot especially in the north-west. Rock Samphire is common in the south and west, often growing on sheer cliff faces, while Scots Lovage is only to be found in Scotland and northern Ireland. The short, springy turf on the cliff tops makes walking a delight, and it is there that most of the flowers will be found. Sea Campion, Sea Pearlwort, Rock Sea-spurrey, and Buck's-horn Plantain are all common, with Sea Stork's-bill, Thyme, and Common Rock-rose in drier, sunny sites, and Common Scurvygrass in damp areas where streams trickle over the cliff edge. In these wetter spots Wild Celery grows, as well as Alexanders in the south of England. In the short turf of the cliff tops in Cornwall, Wales, much of the south-west coast of Scotland, and especially Orkney and Shetland, look for the delightful little Spring Squill, with flowers like blue stars; Sutherland and Orkney cliff tops are also the favoured sites for the Scottish Primrose.

Some botanical terms

Terms mentioned within a definition that have been defined elsewhere are printed in SMALL CAPITALS.

achene a dry fruit containing one seed and and not splitting open
acid water or soil containing free acids with a *p*H value less than 7, lacking in chalk and lime
adventive an alien plant not permanently established
aggregate a group of closely related species, treated as a single species
alkaline water or soil with a *p*H value greater than 7, containing lime or potash
alkaloid a complex organic substance, often highly poisonous
alluvial soil formed of fine particles washed down by rain or rivers and deposited in a valley
annual a plant completing its life cycle in one year
anther the sac at the tip of the stamen which contains the pollen grains
apomixis reproduction by seed not formed as the result of sexual fusion
appressed pressed close, but not joined on
awn a stiff bristle-like projection on a fruit or leaflet
axil the angle between the stem and a leaf stalk.
basic ALKALINE, containing compounds of calcium magnesium or potassium
beak the pointed tip of a seed pod.
biennial a plant taking two years to complete its life cycle
bisexual a flower containing organs of both sexes
brackish freshwater containing some seawater.
bract a modified leaf at the base of a flower stalk, with a flower in the AXIL
bracteole a tiny leaf on a flower stalk wthout a flower in the AXIL
bulb an underground swelling at the base of a plant, formed of overlapping fleshy leaf bases
bulbil a small bulb, usually formed in a leaf AXIL, or on a leaf margin
calcareous containing chalk or lime
calcifuge a plant which avoids soils containing chalk or lime
calyx the whorl of SEPALS, sometimes joined together into a tube
capsule a dry fruit containing several seeds, splitting open to release them
carpel one of the divisions of the CAPSULE
caruncle the humps on either side of the lip of some orchids
chlorophyll the green pigment present in most plants cells
cladode a modified stem, flattened to resemble a leaf
cleistogamy self-pollination within an unopened flower
column a specialized structure in the centre of an orchid flower, bearing the male and female organs
composite a member of the Daisy family
coniferous plants such as pine and juniper, which bear cones
cordate heart shaped
corm a solid swelling of the underground base of a stem, a storage organ
corolla the whorl of PETALS sometimes joined together into a tube
crucifer a member of the Cabbage family
deciduous plants which shed their leaves annually in the autumn
dioecious having male and female flowers on separate plants
dissected finely divided
enzyme a complex organic substance made within a cell, which assists in chemical reactions
epicalyx a whorl of BRACTS or leaflets below the true CALYX
epichile the outer part of the LABELLUM of flowers of the genus *Epipactis*
episepals small SEPALS formed between the true sepals
falls the three broad outer PETALS of an iris, which hang downwards
filament the stalk of the STAMEN which bears the ANTHER
floret a small individual flower. In the Daisy family divided into disc-florets in the flower centre, and showy ray-florets on the periphery
foetid having an unpleasant smell
glandular bearing cells which secrete liquid substances
glaucous bluish, usually smooth
gynodioecious having female and bisexual flowers on separate plants
gynoecium the female parts of a flower, the PISTIL
hermaphrodite flowers containing male and female organs
hip fruit of the rose
hood the helmet shape formed by the upper PETALS and SEPALS in some orchids and other plants
host the plant upon which a PARASITE feeds
humus decaying organic matter in the soil, digested and degraded by earthworms, fungi, and bacteria

hybrid a plant arising from the fertilization of one species by another

hybrid swarm a group of hybrids showing a range of characteristics between those of the two parent plants

hypersensitivity an excessive systemic reaction to a plant protein

hypochile the cup-shaped basal portion of the LABELLUM of orchids belonging to the genus *Epipactis*

involucre a whorl of BRACTS forming a collar below a compound flower head

keel boat-shaped structure in a pea flower formed by the two lower conjoined PETALS

keeled often applied to leaves with a strong midrib, folded like the keel of a boat

labellum the lower PETAL of an orchid flower, often large, complex in structure and showy

labiate a member of the plant family which includes mint and thyme

leached describing soil from which the soluble elements have been removed by water action

linear long and narrow with parallel sides

lip the large lower, and sometimes also the upper, LOBE of the COROLLA

lobe a large division of a leaf or PETAL, which is not separately stalked

machair Hebridean grassland by the sea on a foundation of calcareous sand

microspecies a distinct group of plants within a species, showing small but constant differences in character

midrib the strong central vein of a leaf

monoecious having separate male and female flowers on the same plant

montane associated with mountains

mucilage a sticky or slimy juice obtained from the leaves or roots of a plant

mycorrhiza a fungus which invades the underground tubers or roots of orchids and plants such as heathers

native a plant natural to a country, not introduced

naturalized an introduced plant now totally adapted to its new environment

nectary a structure, often near the base of a PETAL, which secretes nectar

neutral neither ACID nor ALKALINE, having a *p*H value of 7

nitrogenous substances containing nitrogen, often simple compounds from which complex proteins can be built

node that point on a stem, sometimes swollen, where a leaf or group of leaves, is attached

ovary the lower part of the female reproductive organ which contains the seeds, composed of one or more CARPELS

ovule the structure containing a single egg, developing into a seed after fertilization

palate the swollen base of the lower lip of flowers such as toadflax and bladderwort, which closes the mouth of the flower

palmate divided like the fingers of a hand

pappus a ring of hairs, sometimes feathery, around the top of seeds of plants such as Daisy and Dandelion

parasite a plant living at the expense of another, or HOST plant

perennial a plant living for more than two years

perianth the outer, non-reproductive, parts of a flower, divided into inner PETALS (COROLLA) and outer SEPALS (CALYX)

petal one of the inner PERIANTH segments, often brightly coloured

***p*H** a logarithmic measure of the hydrogen ion concentration in moles per litre; values greater than 7 are alkaline, less than 7 are acidic

photosynthesis the process of forming carbohydrates from carbon dioxide and water, using the energy of sunlight and chlorophyll as a catalyst

phyllary specialized BRACTS around the flowers of members of the Compositae (daisies and thistles)

phyllode a modified leaf stalk, flattened to appear like a leaf

pinnate of a compound leaf, the leaflets arranged in opposite pairs on either side of the MIDRIB

pod a dry fruit containing one or more seeds and splitting down two sides, such as fruits of peas and vetches

pollinia structures formed by coherent pollen grains, usually in orchids

potash a salt of potassium

proteolytic an ENZYME which digests protein

pseudocopulation the attempt by a male insect to mate with a flower to which it has been attracted

ray one of the stalks of an UMBEL, or the strap-shaped outer FLORETS of a composite flower

receptacle the tip of the flower stalk to which the SEPALS, PETALS and other structures are attached

relict plants whose present local population represents a small part of what was once more widespread

rhizome a creeping underground stem, sometimes stout as in the iris, and used as a storage organ

rootstock a short underground stem, usually growing vertically

rosette leaves arranged around the stem base like the petals of a rose, often flat on the ground

rostellum the sterile third STIGMA of an orchid flower, situated between the two functional stigmas, often long and beak shaped, bearing the VISCIDIA of the POLLINIA

runner a prostrate creeping and rooting shoot above ground

saprophyte a plant which obtains its nutrition from the breakdown of dead plant or animal material

semipeloric a flower in which abnormal PETALS and SEPALS give a spurious appearance of symmetry

sepal a division of the CALYX – one of the outer whorl of PERIANTH segments

septum a partition

sessile stalkless

spadix a dense erect spike of FLORETS found in flowers of the Arum family

spathe a large BRACT or pair of bracts which envelop a developing flower spike

sphagnum a family of mosses, members of which are characteristic of the flora of bogs

spike an elongated unbranched flower head at the tip of the flower stem

spur an elongated pouch formed at the base of a SEPAL or PETAL

stamen one of the male organs of a flower, having a FILAMENT bearing an ANTHER at the tip

staminode a sterile STAMEN, much altered in shape; present in the Lady's Slipper

standard the broad upper PETAL of a pea flower or the three inner erect petals of an iris

stigma the receptive upper part of the female reproductive organ

stipule a leaf-like appendage at the base of a leaf stalk, often borne in pairs

stolon an above-ground, creeping stem

style the stalk arising from the tip of the OVARY, which bears the STIGMA or stigmas at its tip

symbiotic the relationship between two different organisms living together to their mutual advantage

tendril a curling, climbing organ often formed by modification of the terminal leaflet of a PINNATE leaf

terminal borne at the tip of a stem

trifid split into three

trifoliate a leaf composed of three leaflets

tuber swollen part of a root or underground stem used as a storage organ

turion a detached bud of an aquatic plant by means of which the plant overwinters

turlough a pond or small lake which dries up in summer (Irish)

umbel an umbrella-shaped flower head with branches arising at the same level, often flat topped; in a complex umbel each branch in turn bears a secondary umbel

umbellifer a member of a large family which includes Carrot, Parsley, Hemlock, and Alexanders

unisexual bearing organs of one sex only

variety (abbreviated to var.) a division of a species showing minor differences of character

vegetative concerned with growth and development

viscidium the sticky disc at the base of the POLLINIA of an orchid flower, which glues it to a visiting insect

whorl more than three leaves or flowers arising at the same level on a stem

winged applied to a stem with one or more flanges along its length

wings the side PETALS of a pea flower

xerophyte a plant adapted to growing in dry conditions

zone a distinct area within a habitat, usually the result of the action of physical conditions

The Calendar

JANUARY

The first two months of the year are, without doubt, the coldest, with temperatures often at or below freezing point, and snow may cover the ground, too. Plant growth ceases at such low temperatures but, in the south and south-west, where the climate is less severe, some hardy plants will flower. It is worthwhile keeping an observant eye open when passing by sheltered, south-facing hedges or woodland borders, particularly where there is a slope to catch the infrequent rays of the sun. In such places you may find the out-of-season Primrose. Although March is the proper time for it to flower, you may have the pleasant feeling of cheating winter and shortening the time until spring.

In sheltered parts of the garden, Shepherd's Purse, Common Chickweed and Groundsel will all show the occasional flower, especially if the winter is mild, but pride of place must go to Winter Heliotrope. The scent is remarkably powerful; it will even reach the inside of a passing car with the windows closed, and the almond-like smell is a promise of better days to come. Those who commute into the cities should spare a glance at the railway banks, which have become splendid, undisturbed nature reserves; Winter Heliotrope and Primroses often flourish there. In many gardens the young plants of annual weeds such as Hairy Bittercress appear in droves as soon as the soil has been warmed for a day or two; the name 'Hairy Poppers' describes well their ability to pop up unexpectedly. Petty Spurge is another weed which will flower in the odd warm corner in January although it belongs more to early summer.

January is never really 'dead'; the first stirrings of growth are there for the observant to see. In 1975, *Gagea bohemica,* a close relative of Yellow Star-of-Bethlehem *(Gagea lutea),* was found flowering in mid-January in Radnorshire, the first time that this plant had been recorded in Britain. Ten years previously a faded flower had been collected accidentally with mosses in April 1965 and, since the faded petals were white instead of yellow, it had not been identified. Only by revisiting the site in January was its true nature revealed – an excitement indeed for a 'dead' month!

Shepherd's-purse

Capsella bursa-pastoris (L.) Medic.

30-400mm

Shepherd's-purse is one of the most successful weeds of cultivated and waste land. It is widely distributed throughout Britain and grows in gardens, fields, and roadsides from Cornwall to the Shetland Isles, especially where the ground has been disturbed. On the north-west seaboard of Scotland it is less common and it seldom grows on mountain above 800 metres. The basal leaves form a rosette and the upper leaves clasp the stem which bears an erect cluster of small white, four-petalled flowers. The distinctive seed pods of Shepherd's-purse are heart shaped and flat. The flowers are self-pollinated and set abundant seed, to the chagrin of the keen gardener. This is the key to the success of the species; because such plants are independent of insects for pollination, they were able to colonize rapidly ground exposed by the retreating ice of the last Ice Age. Shepherd's-purse flowers in every month of the year, and it is one of the very few plants which can reliably be found in flower in January. Many of us have school-time memories of drawing sad specimens during winter biology lessons, drawings which somehow never fitted into the piece of paper provided.

Common Chickweed

Stellaria media (L.) Vill.

50-300mm

Common Chickweed is one of the few plants to flower early in January although it may be found in flower throughout the year. It grows in all parts of Britain in cultivated and waste land and, like Shepherd's-purse, it is a highly successful, self-pollinated weed. The plants are weak and straggly with pale-green, heart-shaped leaves which are almost hairless. The rounded stems bear a single line of hairs on alternate sides between the leaf nodes. The flowers have five petals which are so deeply cleft that they appear to be ten. The petals fall very readily from the flowers and the anthers on the stamens are reddish in colour. These features help to distinguish it from the other eight chickweeds and stitchworts to be found in Britain, although Common Chickweed is the only one likely to be found in flower so early in the year. Common Chickweed can be found growing in most gardens and along the edges of cultivated fields. It is a useful source of green feed for cage birds at a time of year when such greenery is scarce and, in the past, it has been used in ointments for the treatment of dermatitis.

Groundsel

Senecio vulgaris L.

A highly successful weed of gardens, arable, and waste land, Groundsel grows throughout Britain. It can be found in flower in all months of the year, those plants flowering in the winter tending to be smaller in stature. One reason for its success is the nature of the seed which is equipped with a parachute-like plume, so that it is easily dispersed by the wind. The fruiting heads of Groundsel form a spherical mass like a tiny Dandelion clock, and anyone who has tried to clear seeding Groundsel from a garden will vouch for the ease with which the seeds spread. The yellow flowers are small and, because they lack the showy ray florets which make up the disc of flowers such as the Dandelion, they are cylindrical in shape. Around the base of the flower are short, pointed bracts, the lower row being tipped with black. Groundsel is so widely distributed that it may be found almost anywhere in disturbed ground but, in January, it is more likely to flower in gardens or on derelict sites in towns which are a few degrees warmer than the open country, growing with Shepherd's-purse and Common Chickweed. The leaves are a useful green feed for cage birds and rabbits.

80-400mm

Winter Heliotrope

Petasites fragrans (Vill.) C. Presl.

Winter Heliotrope is not a native plant in Britain but it is now widely distributed and locally abundant in many areas, especially in south and south-west England and the southern half of Ireland. It is uncommon in much of central northern England and north-west Scotland but grows around Glasgow, Edinburgh, and on the south side of the Moray Firth. The plants soon escaped from gardens, where they were grown for their characteristic vanilla-scented flowers, and are now found in a variety of habitats, particularly road verges and sunny banks. The powerful white rhizome spreads underground with great efficiency, even pushing up through tarmac surfaces; where the plant is well established it is difficult to eradicate. The flowers appear in January, a month before the yellow flowers of Colt's-foot *(Tussilago farfara)* which grows in similar places. The pale-lilac-coloured flowers are similar to those of Butterbur *(Petasites hybridus)* and make an attractive sight beside an otherwise dreary winter roadside. The large kidney-shaped leaves appear at the same time as the flowers. They are evenly toothed and green on both surfaces but they are easily scorched by frost.

100-250mm

FEBRUARY

Early February is often bitterly cold but, by the latter part of the month, the sun is beginning to warm the soil and, even on the western moors, Round-leaved Crowfoot will flower in sheltered ditches and pools.

In south-west England in warm corners above the beaches, and in damp clefts among the rocks, Danish Scurvygrass comes into bloom, the earliest member of the genus *Cochlearia* to flower, often accompanied by Red Dead-nettle. On sunny hedge banks, Barren Strawberry opens its white flowers among the dead leaves and brown remains of the last season, with Dog's Mercury forming bright, yellow-green patches of new leaves right underneath the hedges and carpeting the floor of the woods.

Butcher's Broom may be the only plant flowering in some dry woodlands on acid soil in the south. At this time of year the dark-green, bushy plants are easier to see above the dead leaves. The tiny flowers have six pointed white petals, set like a tiny star on the surface of the dark-green 'leaf'.

In a few deciduous woodlands in the south and east, Winter Aconite flowers are breaking through the bare ground although you are more likely to see them planted in parks and gardens.

The most eagerly awaited February flowers must surely be Snowdrops, at their best in the middle of the month. They have become widely naturalized in gardens, parks, and churchyards, flowering bravely even through the snow. Frost and rain may flatten them but, at the first hint of sun, they recover and look as fresh as ever.

Plants cannot flower so early in the year unless they have a store of energy in a bulb or some other underground, frost-proof structure. Italian Lords-and-ladies has thick, tuberous roots packed with starch-like material from which the flowering plant develops. The unspotted leaves are thicker and blunter than those of the common Lords-and-ladies, and appear above ground much earlier in the year, so that they are fairly well formed in February before the common species has begun to sprout. It is an uncommon plant of sheltered copses from Cornwall to Sussex, and this is an ideal time of year to find it, when little else is growing. Colt's-foot has thick, white rhizomes deep below the ground surface, and the yellow, dandelion-like flowers, with stems sheathed in purplish scales, are a welcome sight on road verges and in waste places everywhere. When Colt's-foot is in bloom spring is beginning.

Winter Aconite

Eranthis hyemalis (L.) Salisb.

70-150mm

Winter Aconite is a charming southern European species grown in gardens as an early spring flower and naturalized widely but infrequently in parks and woods. It grows in the southern lowlands of Britain, particularly in Suffolk, west Norfolk, Cambridgeshire, Northamptonshire, and Leicestershire, and has been recorded as far north as Fife and Kincardine but not in Ireland. In most localities it is rare and seems to be decreasing. The flowering stem bears a solitary yellow flower which appears before the glossy, palmate leaves. The flowers are formed by six yellow sepals below which there is a broad, pale-green ruff of three deeply divided stem leaves. The petals are reduced to minute, scale-like structures containing the nectary, set at the base of the sepals and tucked under the carpels. Winter Aconite should be sought from mid-February onwards in bare ground in sheltered woods and in sites such as churchyards, where there is a moderate amount of leaf mould and not too much competition from larger plants, so that the sun can penetrate to warm the soil. There Winter Aconites may be accompanied by Snowdrops *(Galanthus nivalis)*.

Round-leaved Crowfoot

Ranunculus omiophyllus Ten.

6-12mm

Round-leaved Crowfoot is one of the very few flowers that you are likely to find during a moorland walk in February. Although the main flowering period is early summer, a few can be found blooming in shallow pools and boggy, wet places on peaty soils in many places down the western side of Britain. Round-leaved Crowfoot is recorded from south-west Scotland to Cornwall, especially on the moors of Cumbria, west Lancashire, and west Yorkshire, and on the heaths of southern England. It also grows on the Isle of Man and in southern Ireland. The creeping stems bear kidney-shaped leaves with rounded lobes and flowers resembling small white, yellow-centred buttercups. Round-leaved Crowfoot can be distinguished from the closely related Ivy-leaved Crowfoot *(R. hederaceus)* by its larger flowers with petals that are twice the length of the sepals, and more rounded, less deeply lobed leaves. All members of the crowfoot family are rather delicate, lacking the coarseness of buttercups, and often make a spectacular sheet of white flowers over the surface of pools and ditches. The dozen species which grow in Britain show considerable variation in form, depending upon the habitat in which they occur.

Barren Strawberry

Potentilla sterilis (L.) Garcke

Barren Strawberry bears a superficial resemblance to Wild and to Cultivated Strawberries, spreading, as they do, by runners which root at intervals, but its fruits are quite unstrawberry-like. The receptacle does not swell to form the familiar strawberry but is covered by a mass of dry, hairy achenes. Barren Strawberry flowers from February to May which is much earlier than Wild Strawberry *(Fragaria vesca)*. The flowers can be distinguished by the distinct gaps between the five white petals which are themselves slightly notched. There are five episepals which lie between the main sepals and form a cup beneath the fruit. These are almost as long as the sepals, while those of *Fragaria* species are longer. The lustreless leaves have less strongly marked veins, and the terminal tooth at the end of each leaf is shorter than its neighbours. Barren Strawberry is widespread in dry grassy places such as road verges, the floors of abandoned quarries, and beside footpaths. You can also find it in hedgerows and on the edges of woods throughout Britain, except for the Outer Hebrides, Orkney, and Shetland. It is a common plant in southern England but is less frequent than Wild Strawberry on chalky soils.

50-130mm

Dog's Mercury

Mercurialis perennis L.

Dog's Mercury is widespread and common as a plant of woodland and shady hedge banks in most of England and Wales. The stiff, unbranched stem springs from a slender, creeping stolon, and bears dark-green, pointed, shallow-toothed leaves in pairs on long stalks. The insignificant pale-green flowers are carried in loose clusters on long stalks, male and female on different plants. The male flowers have three tiny green perianth segments and a cluster of stamens, while the female flowers, which appear some time after the male, have two styles which persist above the two round, hairy carpels. Each carpel contains a single large seed and is grooved on the outside so that there appear to be four. Dog's Mercury is not native in Ireland, except in County Clare, and is absent from the Isle of Man, the Outer Hebrides, Orkney, and Shetland. It is very rare in the north of Scotland in Ross, Sutherland, and Caithness. In southern woodlands it often grows in a dense carpet excluding all other competitors while, in much of the north, it grows on damp cliff ledges and between boulders by the sea and in the mountains. The foliage is poisonous to cattle, causing a severe enteritis and slimy yellow milk.

150-400mm

Red Dead-nettle

Lamium purpureum L.

100-300mm

Although it is one of the less showy members of the Thyme family, Labiatae, Red Dead-nettle is not without charm, especially when the small, purple-red flowers and downy leaves are silvered with frost, as they may often be in February. It is abundant in waste and cultivated ground throughout Britain, although it is less common in Wales, west Scotland, and Ireland. Red Dead-nettle flowers from early February onwards and, in coastal areas in the west, it grows in rough ground behind beaches with Early Scurvygrass *(Cochlearia danica)*. The whole plant of Red Dead-nettle is downy and has a pungent smell when crushed. The pointed, heart-shaped leaves, sometimes tinged with purple, have shallow, coarse teeth and are all borne in stalked pairs up the stem. The reddish-purple flowers are carried in a cluster at the top of the stem, each set in a hairy calyx; the lower lip of the flower is forked and marked with small, dark-red spots. Two closely related species flower later in the spring. Cut-leaved Dead-nettle *(L. hybridum)* is more slender, with deeply dissected leaves, while Henbit Dead-nettle *(L. amplexicaule)* has unstalked upper leaves and more prominent flowers in several whorls down the stem.

Colt's-foot

Tussilago farfara L.

50-200mm

Colt's-foot is one of the welcome harbingers of spring, and a few warm days in February will suffice to bring out its bright-yellow flowers. Like so many early flowering perennials, it has a stout underground rhizome, so that it is relatively independent of the weather and can use a store of food laid down in the previous summer. The flower stems appear before the leaves and are sheathed in overlapping, purplish, woolly scales. The erect yellow flowers have both disc and ray florets, unlike the Dandelion *(Taraxacum officinale)*. They droop as soon as they have finished flowering but become erect again in fruit. The seeds have a long silky plume, the whole mass forming a large spherical 'Dandelion clock'. The leaves appear well after the flowering period. They are broadly heart shaped with pointed teeth, downy green above and grey felted on the underside. Colt's-foot can be found throughout Britain in waste ground, sand-dunes, and riverside shingles, and shows a liking for clay soils. Although it usually flowers in February and March, it can be found flowering in June and even later in mountains where the snow has only just melted.

Butcher's-broom

Ruscus aculeatus L.

Butcher's-broom is not infrequent in dry woodlands, especially in southern England in the oakwoods of the New Forest and the Sussex Weald, and in East Anglia, south-west England, and west Wales. It is a most curious plant and, although it is a member of the Lily family, Liliaceae, it is evergreen, tough, and woody, bearing what appear to be sharply pointed, dark-green leaves. These structures are, in fact, modified stems or cladodes, while the true leaves are reduced to tiny chaffy scales. The greenish-white flowers, which appear in February, are very small and carried in the middle of the upper surface of the cladodes. They are unisexual, male and female flowering on different plants and, despite their diminutive size, they are rather pretty and well worth examining through a hand lens. The fruit is a round red berry, ripening in the autumn and remaining on the plant during the winter. The dark-green foliage with the contrasting red berries is sometimes picked for winter decoration, a prickly job, and, because the berries are easily dislodged, artificial coloured berries are sometimes stuck on in their place. The resulting concoction used to be sold by pedlars at the door.

230-800mm

Snowdrop

Galanthus nivalis L.

The first Snowdrop of spring is something to be anticipated with pleasure each new year. Snowdrops will flower early in February, even pushing up through a crust of snow. It is small wonder that they are planted in gardens all over the country and are now naturalized in woods, copses, and damp meadows. They may be native in parts of Wales and western England where sheets of flowers grow beside some Devon rivers. They become less common further north in Scotland and are absent from Orkney, Shetland, and Ireland. The two bluish-green leaves appear above the ground first, followed by the single drooping flower. The three outer 'petals' are long, white, and spreading while the three inner 'petals' are shorter, notched, and marked with green at the tip. Snowdrop bulbs have caused poisoning when fed accidentally to livestock but, under normal circumstances, the foliage is not eaten. A similar species, Spring Snowflake *(Leucojum vernum)*, is also grown in gardens but, in the wild, it is very rare, known only from two small areas in south-west England. It flowers later than Snowdrop, in March and April, and has fatter, more bell-shaped flowers.

100-250mm

MARCH

When the weather permits, this is a good time to walk through the lowland woods and along country lanes, even if it is wet and muddy elsewhere. Under hedges and on banks Primroses are now coming out, and Sweet Violet and Early Dog-violet make splendid patches of colour. Even the foliage of Sweet Violet has a delicious fragrance when disturbed, and plants with white flowers are not uncommon and seem especially fragrant. The Butterburs will be pushing up their fat clusters of buds, especially on road verges, the broad leaves opening out later in the spring. In the west country and in Pembrokeshire, the lanes at this time of year are particularly fresh and lovely, and walking is a real pleasure without the hazard of the summer traffic.

In the woods, Lesser Celandine carpets damp areas and beside streams, while Wood Anemones are dominant in the drier parts. It is delightful to see a carpet of Wood Anemones, and it is always worth looking for unusually coloured flowers, flushed with pink or, rarely, pale blue. Wood Spurge and Spurge Laurel are in flower beneath the trees in woods on chalky soils, with Stinking Hellebore and Green Hellebore or Bear's-foot in a few deciduous lowland woods in the south of Britain.

Where woodland streams run through deep gullies, the moist, mossy sides may be a site for Opposite-leaved Golden-saxifrage. Wild Daffodil flowers in copses and woods in many areas of the south and south-west; they bloom a little later in the Lakeland sites immortalized by Wordsworth.

In waste ground, in gardens, and on lawns, many of the speedwells are coming into flower; Wall Speedwell, Ivy Speedwell, and several other common species may bloom while, in the Breckland of East Anglia, the rare Breckland Speedwell can be found.

In marshy areas and beside streams, Marsh-marigold makes a mass of deep-yellow blooms, with Bulbous Buttercup in the fields nearby. In gardens, Hairy Bittercress is flowering, soon to set seed and plague the unwary gardener with another crop of plants to weed out in early summer.

The end of the month is the time to look for Wallflower on chalk cliffs in the south and west and, in one part of east Sussex, it has been spread by roadworks from its old cliff site to verges and roundabouts on a new bypass.

On the moors and mountain tops, Crowberry will be coming into bloom and, in western Ireland, this is the time to find Irish Heath in flower. In March everything is beginning to stir and patches of bright new growth are appearing everywhere. There is a freshness in the air; spring is here.

Marsh-marigold

Caltha palustris L.

This impressive plant is known by many people as Kingcup although it has many names including the delightful one of Mollyblobs. Marsh-marigolds are found throughout Britain in marshes, ditches, on the edges of pools, and along the banks of streams, especially where there is deep, rich mud. In the north they will be flowering well into June, making a splendid splash of golden yellow in such diverse places as the streams below Kilnsey Crag in Yorkshire, the ditches on the machair of Tiree in the Hebrides, and even high up at 800 metres on Ben More in Sutherland. The lowland form can be massive with large, toothed, kidney-shaped leaves which are often mottled, and large, yellow, buttercup-like flowers formed by five sepals (there are no petals). The mountain form is smaller with creeping stems and more triangular leaves. Marsh-marigolds are among the earliest waterside flowers to bloom in the south; they bring back childhood memories of the curious sickly sweet smell of their crushed foliage. They contain a poison called protoanemonin, which is destroyed by drying in hay, but cattle eating the fresh plant salivate, and suffer colic and dysentery from the blistering effect of the poison.

150-450mm

Wood Anemone

Anemone nemorosa L.

The flowering of Wood Anemones ushers in the spring, and there is no lovelier sight than a leafless wood carpeted with myriads of the white, star-like flowers. Wood Anemone flowers are solitary on a slender stem above a whorl of three divided, trifoliate leaves, with more stalked leaves below. The flowers have no petals but are formed by five or more hairy white sepals, sometimes suffused with pink on their outer surfaces, around a cluster of yellow stamens. One unusual colour form has flowers of clear ice blue and these are particularly fine. Wood Anemones are often abundant in deciduous and mixed woodlands and under hedges, wherever the soil is not too acid, flowering with Sweet Violet *(Viola odorata)*, Early Dog-violet *(V. reichenbachiana)*, and Primroses *(Primula vulgaris)*. They also grow in sheltered gullies and on cliff ledges quite high up on mountains in Wales and Scotland, although they are absent from Orkney, Shetland, and much of the Hebrides, and are rather uncommon in the south of Ireland. Wood Anemones, or Wind Flowers as they are also called, have a delicate sweet scent.

2

60-300mm

Lesser Celandine

Ranunculus ficaria L.

Lesser Celandine is another widespread and abundant flower of early spring, enjoying rather wetter conditions than its near relative the Wood Anemone. You will find its shining yellow flowers in damp woods and on shady banks, often beside water. The long-stalked leaves are heart shaped, dark green, and shiny, springing in a floppy rosette from a clump of fat dumpy roots. The solitary flowers are also shiny, with eight to twelve pointed, glossy petals forming a bright-yellow star. The flowers open wide in sunshine, but will close up in dull weather revealing the three green sepals underneath. Look carefully in the leaf axils where some plants will form reproductive bulbs. Such individuals often have infertile fruits and the bulbs are a means of vegetative multiplication. Lesser Celandine is a very common plant throughout much of Britain; in parts of north-west Scotland and in Ireland it is less abundant. In the damp woods where it grows in southern England it is always worth looking for the two rare hellebores, the Bear's-foot or Green Hellebore *(Helleborus viridis)* and Stinking Hellebore *(H. foetidus)* with more bell-shaped flowers edged with purple.

50–250mm

Wallflower

Cheiranthus cheiri L.

Wallflowers were probably introduced into Britain from south-east Europe but they now grow in widely scattered areas of Britain and Ireland as far north as the Moray Firth in Scotland. Although you will find them growing on old crumbling walls, this is probably because of the lime mortar used in building, and they are more naturally at home on cliffs and on quarry faces especially where the rock is chalk or limestone. They are most common along the coast of Kent and Sussex in such places as the cliffs around Beachy Head, on the coast of Devon, around the Severn estuary, in Wales on the coast between Conwy and Prestatyn, and in north Norfolk. Wild Wallflowers differ slightly from the garden wallflowers which are derived from the genus *Erysimum*. They are biennial, but can become perennial with a tough, woody stem. The flower buds are enclosed by brown sepals, from which the orange-yellow petals emerge; the flowers are strongly scented especially on sunny days. On the dry walls and cliffs where Wallflowers grow, you will also find Common Whitlowgrass *(Erophila verna)* with its small white flowers and curiously shaped seed pods.

200–600mm

Heath Dog-violet

Viola canina L.

I first found Heath Dog-violet in flower among the sand-dunes of Morfa Harlech on a lovely spring day. The clumps of blue flowers mirrored the blue of the sea and sky. It is one of our most attractive violets, the colour of the petals contrasting with the bright yellow of the spur. Heath Dog-violet can be found throughout Britain and Ireland on heaths, dry grassland, dunes, and, more rarely, in some fens. Although it grows abundantly in sandy coastal areas as far north as north-west Sutherland, Orkney, and Shetland, it is more frequent in the southern half of England. There it also occurs in patches of sandy leached soil overlying the chalk downs near the sea. The leaves are dark green, pointed, often a little leathery, with a shiny upper surface, and small teeth. The sepals are pointed and the petals bright blue, contrasting with the stout straight spur which is yellow and bears a shallow notch at the tip. Heath Dog-violet flowers as early as March in southern England, while far to the north in Shetland it will be at its best in May.

20-250mm

Opposite-leaved Golden-saxifrage

Chrysosplenium oppositifolium L.

Opposite-leaved Golden-saxifrage is a lover of wet places, forming a mat of green leaves and yellow flowers in damp woods and on the banks of streams and ditches. In mountainous areas you can find it at the foot of rock faces where there is a constant flow of water. The mat-forming stems creep and root, bearing rounded, bluntly toothed leaves in opposite pairs, as distinct from the long-stalked, kidney-shaped leaves of the much rarer Alternate-leaved Golden-saxifrage *(C. alternifolium)*. The flowers, formed by four golden sepals, are borne in flat clusters backed by large, yellow-green bracts. This gives them the appearance of little golden eggs on a green plate. Opposite-leaved Golden-saxifrage is widespread and locally common throughout Britain except in the drier eastern counties of England. It prefers wet, shady places on acid soils and, in the south, it grows away from the coast in woodlands where streams have cut deep gullies down through the rock. Alternate-leaved Golden-saxifrage is more robust and has large flowers. It does not form such dense mats and, where the two saxifrages grow together, it tends to grow deeper in among the rocks. It is absent from the extreme west and grows on more basic soils.

50-150mm

Mezereon

Daphne mezereum L.

Mezereon is widely cultivated for its attractive pinkish-purple flowers borne in clusters on the bare stems before the leaves open. The leaves of Wild Mezereon, however, often appear with the flowers which are paler and less numerous. Mezereon is a very rare plant of woods on chalk and limestone. It is known in a few sites in southern England from Wiltshire to east Sussex, in the Chilterns, and in the north on the limestone of west Yorkshire, Staffordshire, Derby, and Cumbria. The leaves of Mezereon are thin, pale green, and easily mistaken for recently opened leaves of Privet *(Ligustrum vulgare)*. The purplish flowers are tubular, with four spreading lobes formed by the fused sepals, and have a pleasant sweet scent. The fruit is a shiny scarlet berry, set close to the stem, and containing an acrid, irritant poison. Animals will rarely touch the plant, but children have been poisoned by mistaking the berries for red currants. Although Mezereon is a woodland plant, it needs a reasonable amount of light to thrive. The reduction of manpower on many estates has led to coppices and rides becoming overgrown so that Mezereon has gone from many of its old haunts.

0·5–1·0m

Spurge-laurel

Daphne laureola L.

Despite its name, Spurge-laurel is related neither to the spurges nor to Laurel although the long, pointed, evergreen leaves bear a faint resemblance to those of Laurel. The tall bare stems are marked with the scars of old leaf attachments and carry top-heavy clusters of drooping, dark-green leaves which are thick and shiny. The four-lobed tubular flowers are formed by the conjoined sepals, in the same way as the flowers of Mezereon, but they are smaller, less strongly scented, and altogether less showy. They are pale greenish yellow and grow in substantial clusters partially hidden under the upper leaves. Spurge-laurel is found most commonly in beechwoods on chalk in south and south-east England, although it occurs in a few places in south-west England and Wales. It is recorded as far north as Cumbria in woods on limestone, and rarely in Argyll. Spurge-laurel will tolerate a considerable amount of shade and, in beechwoods, it may be the only plant, apart from the rare Bird's-nest Orchid *(Neottia nidus-avis)* or Lady Orchid *(Orchis purpurea)*, to grow in the bare ground under the trees. The black berries contain the same type of poison as those of Mezereon.

0·3–1·0m

Wood Spurge

Euphorbia amygdaloides L.

Wood Spurge is a common plant of damp woods and hedges on basic or neutral soils in southern England. It is absent from counties north of a line from the Wash to the Dee, rare in west Wales, and found in Ireland only in mid-Cork. A notable feature of downland woods in March, the Wood Spurge is a robust plant with floppy green leaves crowded half-way up the stem and topped by clusters of bright yellow-green flowers. The downy stems are often tinged with red. The flowering head arises from a ruff of leaves and is made up of five to seven spreading rays, each bearing a 'flower', set in a saucer formed by two rounded, yellow-green bracts joined together. The central part, lacking petals and sepals, comprises a cup-shaped structure inside which is a cluster of tiny stamens. The globular female flower topped with three styles projects from this on a stalk and, round the rim of the cup, are set four glands, each shaped like a pair of incurved cow's horns. Each spurge has glands of a different shape, a useful feature in differentiating between species. In common with the other spurges, Wood Spurge contains an irritant milky juice which exudes from cut stems or leaves.

200-800mm

Irish Heath

Erica erigena R Ross

Irish Heath flowers much earlier in the year than any of the other seven heaths and heathers to be found in the British Isles. While they are in flower in the late summer and autumn, Irish Heath blooms from March until May. Like the other rare heaths, St Dabeoc's Heath *(Daboecia cantabrica)* and Mackay's Heath *(Erica mackaiana),* Irish Heath is restricted entirely to the west of Ireland, growing on wet heaths and in bogs in the western parts of Mayo and Galway. There it can be locally abundant. In Europe it grows in Portugal and south-west France, and, like Spring Gentian *(Gentiana verna),* Dense-flowered Orchid *(Neotinea intacta),* and Shrubby Cinquefoil *(Potentilla fructicosa),* it is an example of a plant with a mainly southern European distribution which also favours the west of Ireland. Irish Heath is tall and handsome, reaching 2 metres, with a branched and woody stem. The leaves are smooth and grow in whorls of four up the stem. The flowers are borne in a dense, often one-sided spike which is leafy at the top. They are pale pink, narrow, and tubular in shape, with the reddish-purple stamens half protruding from the mouth of the flower.

0·15-2·0m

Crowberry

Empetrum nigrum
and *E. hermaphroditum* Hagerup

Both species of crowberry are important colonizers of leached, acid soils where they form a dense, shrubby carpet. *E. nigrum* is common on moors in central and north Wales, northern England, and Scotland. It is commoner in the northern half of Ireland than in the south, and also occurs in a few areas of Dartmoor and Exmoor. *E. hermaphroditum* grows at higher altitudes in north Wales on Cader Idris and Snowdon, in Cumbria, and in Scotland on mountains north of the line between Glasgow and Perth, where it may be locally abundant and displace *E. nigrum*. It is absent from Ireland. Crowberry is the major element of crowberry heath, a characteristic zone of dwarf shrubs, including Bilberry *(Vaccinium myrtillus)* and Cowberry *(V. vitis-idaea)*, growing on screes and bare mountain summits from 700 to 1200 metres. Both crowberries have black berries and yellow-green leaves which are shorter and stubbier than those of heather. *E. nigrum* is dioecious, the tiny, pink male flowers being carried on different plants from the female. *E. hermaphroditum* has bisexual flowers and stubbier leaves. Crowberry is eaten by Ptarmigan *(Lagopus mutus)* and the birds' droppings contain easily identifiable leaf remains and resemble heaps of discarded cigarette ends.

50-300mm

Primrose

Primula vulgaris Huds.

Primroses are among the commonest and best loved of all the spring flowers, occurring throughout Britain in woods and on banks, especially beside railway lines. They thrive best on heavy soils. They are also to be found on sea cliffs and in gullies on hills and mountains up to 700 metres while, in the Outer Hebrides where much of the soil is covered by acid peat, they can be found on the edges of eroding streams with other plants more typical of woodland. The name Primrose derives from *Prima Rosa*, the first rose. Although they flower in March, or even during the winter in southern England, Primroses are at their best on the cliffs and by the streams of Orkney and Shetland in May and June, hence their Orcadian name of May Flowers. The fragrant, pale-yellow flowers with darker centres have five notched petals, and are usually carried singly on hairy stalks in a cluster above the rosette of leaves. Pin-eyed flowers have a long style reaching the mouth of the corolla tube, the stamens being attached low down out of sight. Thrum-eyed flowers have a short style which is hidden by the stamens attached around the outer rim of the corolla tube.

30-100mm

Wall Speedwell

Veronica arvensis L.

30-230mm

The speedwells *(Veronica* spp) comprise a large family with confusingly similar flowers. It is essential to use a hand-lens and a good floral key when you are trying to distinguish between the species. There are three main groups. The first group, to which Common Field-speedwell *(V. persica)* belongs, has solitary flowers growing from the leaf axils. The second group has spikes of flowers growing from the leaf axils while the top of the main stem bears leaves only – Brooklime *(V. beccabunga),* which grows in streams, ponds, and marshes belongs to this group. Members of the third group, which includes Wall Speedwell, have flowers in leafy heads on the main stem. Wall Speedwell is very common in fields, wasteland, and on walls throughout Britain. It is small and erect, with pointed-oval, coarsely toothed leaves. The whole plant is downy and the leafy, untoothed bracts tend to hide the flowers because of their short stalks. The corolla is bright blue with a white centre and is shorter than the hairy, pointed sepals. The shape of the seed capsule is an important feature for distinguishing between speedwells. Wall Speedwell has a flattened, heart-shaped capsule as broad as it is long, capped by a short style.

Breckland Speedwell

Veronica praecox Allioni

30-70mm

Breckland Speedwell is a very rare plant now confined to a few sandy fields around Icklingham and Lakenheath in west Suffolk. The Breckland has been reduced to a fraction of the area it formerly covered because of the inroads of arable farming, forestry, and the provision of airfields. Of the three rare speedwells which occur in the Breckland, Breckland Speedwell is usually the earliest to flower. It is a slender, erect annual with unstalked, rounded leaves which have six to ten large, rounded teeth and are often tinged purple on the under surfaces. The flowers are borne on stalks longer than the bracts. The corolla is deep blue, longer than the pointed calyx lobes, and, when formed, the capsule also projects beyond the calyx. The capsule is longer than it is broad and bears a prominent style. Fingered Speedwell *(V. triphyllos)* also has long-stalked, deep-blue flowers, but the leaves have finger-like lobes. Spring Speedwell *(V. verna),* the other rare Breckland species, has flowers with very short stalks and more deeply lobed leaves.

Butterbur

Petasites hybridus (L.) Gaertn. Mey. and Scherb.

Butterbur is a robust perennial, the stout-stalked flowers appearing before the huge, heart-shaped, toothed leaves which measure up to 1 metre across. It is locally abundant in woods and damp places beside streams except in northern Scotland. The unscented male and female flowers are carried on different plants, each composed of small, tubular, five-toothed florets, making a bell-shaped head in the male and a smaller, egg-shaped head in the female. In much of the country only male plants occur, except in the north-west where female plants are more common. Dr Max Walters writes that the male and female plants behave almost as separate 'species', with different tolerances of climatic and soil conditions. Two related species have escaped from gardens or have been introduced into the wild. Giant Butterbur *(P. japonicus)* has creamy coloured flowers and broad, leaf-like bracts overlapping each other up the stem, so that the newly opened flowers resemble a Victorian posy. The smaller White Butterbur *(P. albus),* with scented, white or lilac flowers, is to be found in eastern Scotland but rarely elsewhere.

100–400mm

Yellow Star-of-Bethlehem

Gagea lutea (L.) Ker-Gawl.

This charming little flower is one of a number of stars-of-Bethlehem which grow in Europe and have been grown in gardens, so that it is sometimes difficult to decide which of the colonies in Britain are wild and which have established themselves from introduced plants. Yellow Star-of-Bethlehem grows in damp, deciduous woodlands on basic soils from Dorset northwards through the centre of England, to Angus and Lanark in Scotland, and in a few sites in East Anglia. It is a rare plant which can be locally abundant. The yellow flowers have six pointed perianth members which, in sunshine, open out into a star shape, and bear a superficial resemblance to the flowers of Lesser Celandine. The back of each petal is marked with a green band, and the flowers are carried in a loose, few-flowered umbel on stems 10 to 20 centimetres high. Below the flowers are two green, leaf-like bracts of unequal size, and there is a single narrow, shining leaf which has three to five prominent veins on the underside. Yellow Star-of-Bethlehem rarely sets seed but reproduces by small bulbs formed around the base of the main bulb.

80–200mm

APRIL

The showers and sunshine of April bring a burst of growth and colour to hedges and woods alike. Along the roadsides the ditches are full of Cuckooflower, Garlic Mustard, Greater Stitchwort, and the occasional early Red Campion, but Bluebells are the main attraction, common enough by roads and under hedges, but best seen as a misty carpet beneath the woodland trees which are just coming into leaf. Among them will stand out the occasional purple spike of Early-purple Orchid while, in damper places, there may be a mass of the white-flowered Ramsons, stinking of garlic.

Dry hedges and roadsides will have Herb-robert, Common Dog-violet, Cow Parsley, and Lords-and-ladies, with banks of yellow-flowered Alexanders near the coast in southern England. Parasitic Toothwort and the little symmetrical green flowers of Moschatel are among the many delights of hazel coppice while, along the sunny road verges and in the pastures, Dandelions make a blaze of yellow.

The first of the downland orchids, the Early Spider-orchid, appears at the end of the month at a time when Cowslips are at their best, and this is the season to look for the gracious but local Oxlip. Cowslips have become far less common in much of their range, not only because of changes in farming methods, but as the result of selfish picking by visitors to the country. It does not take that many people picking 'just a few' to decimate what appeared to be a carpet of flowers – they should be left for others to enjoy as well!

Those who live near the limestone hills will find the spectacular Purple Saxifrage at its best while, in Teesdale and the Burren, Spring Gentian is starting to flower, a focus of botanical pilgrimage for those fortunate enough to see them. In the north and west of Scotland the rare Limestone Bugle is in flower. Hemlock Water-dropwort fills the wet ditches with bright-green leaves and flowers, while anyone holidaying by boat on the Thames should look out for the graceful flowers of Loddon Lily by the waterside, like clusters of long-stalked Snowdrops.

Spring Squill is the delight to be sought on cliff tops by the sea in Cornwall and Pembrokeshire, with Common Scurvygrass in the damp flushes where streams run out over the cliffs. At this time of year the old flint walls of the villages near my home in Sussex are brightened by the purple flowers of Ivy-leaved Toadflax, with White Comfrey and Green Alkanet growing at the foot of the walls.

Common Scurvygrass

Cochlearia officinalis L.

50-100mm

The leaves of scurvygrass contain vitamin C, and it is said that they used to be eaten by sailors to prevent scurvy. Common Scurvygrass grows all around the coast of Britain, especially in the west and north, but it is absent from the south-east of England except for a recently discovered site at Bexhill, Sussex where it may have been introduced. It grows in abundance in damp and sheltered places near the sea. It is a robust plant up to 50 centimetres tall, with long-stalked, kidney-shaped, fleshy basal leaves, and more pointed, toothed upper leaves which clasp the stem. The flowers have four white petals, which are twice as long as the rounded sepals. The seed pods are almost spherical, with the pointed remnants of the style at the top, and are carried on prominent stalks. They are often fully formed at the bottom of the flower head while the flowers at the top are still in bud. The identification of the various scurvygrasses is not easy because they are variable in form and tend to hybridize. Alpine Scurvygrass *(C. alpina),* which is widespread in wet gullies especially on the mica-schist mountains of Perth and Angus, may not be a separate species but an inland form of *C. officinalis.*

Lady's Smock or Cuckooflower

Cardamine pratensis L.

250-600mm

Lady's Smock is widespread and common throughout Britain in moist meadows and by the sides of streams, in damp ditches beside roads, and also on mountain ledges, coming into bloom at the time the Cuckoo usually arrives. Carried on stems up to 60 centimetres high, the flowers are large, 12 to 20 millimetres in diameter, and have four fragile lilac petals with darker veins. The sepals are short and narrow and the anthers yellow. The fruit is a long, unbeaked pod on an erect stalk, the pod splitting on either side to expose a row of seeds on a central septum. Lady's Smock is one of the feed plants of the charming Orange-tip Butterfly *(Anthocharis cardamines);* its caterpillars feed on the developing seed pods, not on the leaves. One of the pleasures of a sunny spring day is to see a roadside verge full of the lilac flowers of Lady's Smock, with the little white-and-orange butterflies flitting around them. The closely related Coralroot *(C. bulbifera)* is a rare plant of woods in the Weald and Chilterns. The flowers are darker pink, the pinnate leaflets narrower and slightly toothed while, in the leaf axils, there are dark purplish bulbils.

Garlic Mustard

Alliaria petiolata (Bieb.) Cavara and Grande

Garlic Mustard, or Jack-by-the-hedge, is a very common plant of hedges and banks throughout lowland Britain, especially on the chalk from Kent to Wiltshire. It is uncommon in the north and is absent from the western isles, Orkney and Shetland. It has a scattered distribution in Ireland. It is an annual plant of moderate size, with an erect, little-branched stem up to 120 centimetres high. The leaves are all stalked, heart shaped, and coarsely toothed, smelling of garlic when crushed, a characteristic which gives the plant its name. The flowers are typical crucifer flowers, with four white petals 6 millimetres across and erect, smooth sepals. The seed pods, borne on short, stout stalks are up to 60 millimetres long, containing black cylindrical seeds. The pods of the lower flowers often project through the flowerhead. The developing pods are eaten by caterpillars of the Orange-tip Butterfly although they appear to prefer those of Lady's Smock if the two plants are growing together. Despite the insignificant flowers, the pale-green leaves of Garlic Mustard are a prominent feature of the April hedgerows.

0·2–1·0m

Common Dog-violet

Viola riviniana (Reichb.)

Common Dog-violet is our commonest violet, growing throughout Britain except in areas with alluvial soils such as the levels of the Sussex river valleys and the country bordering the Wash. It often forms extensive patches, a welcome splash of colour on banks, in hedges and shady places, and even in the mountains. It is a perennial with a central barren rosette of leaves and several floppy stems from which the stem leaves and flower stems arise. The leaves which appear in summer are larger than those of spring but both types have a pair of stipules at the leaf base which are fringed with short teeth. The flowers of Common Dog-violet are unscented. The sepals are narrow and pointed, as they are in all the dog-violets. The blue-violet petals are broad, the base of the lower petal prettily marked with dark veins and bearing an upcurved pale spur which is strongly notched at the tip. The spur of the rather similar Wood Dog-violet *(V. reichenbachiana)* is dark, straight, and unnotched. The frequency with which hybridization occurs betwen the dog-violets makes their accurate identification difficult.

20–180mm

Greater Stitchwort

Stellaria holostea L.

The white flowers of Greater Stitchwort, mixed with the mauve of Lady's Smock and the blue of Bluebells, form one of the most attractive sights of ditches and hedgerows in spring. Greater Stitchwort is common throughout most of Britain, except for the area around the Wash and in western Ireland, but is absent from the Outer Hebrides, Orkney, and Shetland. It obviously thrives in moist, warm surroundings such as sheltered ditches, the rather straggly, 60-centimetre-high stems finding mutual support from other plants. You will not find it where the soil is very acid. The flowers are large, 20 to 30 millimetres across, with five brilliant-white petals that are cleft to half-way and are longer than the sepals. This feature is one which helps to differentiate Greater Stitchwort from the other stitchworts, and from the closely related mouse-ears. The stem is rough and four angled. The long, pointed leaves are borne in sessile pairs and are also rough on their margins. The large white petals make Greater Stitchwort as handsome as the other members of the pink family, to which it belongs. Each flower bears a prominent cluster of ten yellow stamens with three styles at their centre; most of the mouse-ears, however, have five styles.

150–600mm

Herb-robert

Geranium robertianum L.

Herb-robert grows throughout Britain in woods, on banks, on shingle, and among rocks on mountains as high as 800 metres. The form will vary depending upon the moistness of the environment, ranging from 50 centimetres tall in sheltered woods to prostrate dwarfed forms on seaside shingle. The lower leaves are divided into five lobes, like the fingers of a hand, and each division is itself deeply divided so that the leaf appears fern-like. The upper leaves are usually divided into three lobes. The leaves and stalks are hairy and are frequently tinged with red, especially on the shingle form where the plants are exposed to high winds, dryness, and a salt-laden environment. The plants have an unpleasant, rank, geranium scent. The bright-pink flowers are carried in pairs on long stalks. The petals are rounded at the tips and are much longer than the pointed, hairy sepals. The pollen is orange. The fruit has an untwisted beak 10 to 20 millimetres long. Herb-robert thrives in soil with a high nitrogen content. In mountainous areas it grows well, with Nettles *(Urtica dioica)* and Great Woodrush *(Luzula sylvatica)*, on ledges and under overhanging rocks where sheep and deer shelter and their droppings accumulate.

100–400mm

Bush Vetch

Vicia sepium L.

Bush Vetch and Tufted Vetch *(V. cracca)* are the two most common hedgerow vetches and occur throughout Britain. Bush Vetch is absent from the highest mountain areas of Scotland although, in the Alps, it can be found up to 2000 metres. Bush Vetch flowers earlier than Tufted Vetch. The rounded, less showy heads of two to six reddish-purple flowers are borne on short stalks, and the plant never scrambles with the abandon of Tufted Vetch with its long showy heads of bluer flowers. Both species have the rounded stems typical of vetches and lack the winged or angled stems of the peas (*Lathyrus* spp). The leaves of Bush Vetch are divided into three to nine pairs of oval, blunt-tipped leaflets like the rungs of a ladder. At the tip of each leaf is a branched tendril, curled like a coiled spring, by which the plant grasps and climbs. It is fascinating to note the speed with which the tendrils will wrap themselves around the twigs and leaves of any plant they touch. Bush Vetch flowers throughout the summer. The seed pods are hairless and black, and have a curled beak at the free end. Children should be dissuaded from gathering the little pea-like seeds because they are not edible and the seeds of several of the *Lathyrus* species are known to be poisonous to children and to horses.

to 1m

Meadow Saxifrage

Saxifraga granulata L.

Meadow Saxifrage is by far the tallest of our native saxifrages, up to 50 centimetres high, and has the largest flowers. It grows on basic and neutral grassland and in sandy places near the sea, mainly in the eastern half of Britain. It occurs as far west as Somerset, rarely in Wales, and, in Scotland, it is more commonly found in the south and east. In Ireland it is known rarely in Wicklow. Unfortunately, it appears to be declining everywhere. One of the finest colonies I have seen was flowering with a mass of Early-purple Orchids *(Orchis mascula)* on the side of a railway cutting where the basic rock was exposed. On the Swedish island of Gotland it is a characteristic plant of the "vicars' meadows" – *prästängst* – which are now nature reserves rich in flowers. The scientific species' name, *granulata,* derives from the cluster of brown, nut-like bulbils which forms in the axils of the basal leaves. These bulbils allow the plant to overwinter safely and propagate vegetatively. The kidney-shaped basal leaves form a rosette from which the tall stem arises, carrying up to twelve flowers. The flowers are large with five well-spaced petals and, when the petals have fallen, the two united carpels become prominent bearing long, divergent horns often tinged with red.

100-500mm

Purple Saxifrage

Saxifraga oppositifolia L.

Purple Saxifrage is a truly arctic-alpine flower. Remains of it have been found in 20 000-year-old glacial deposits in the Lea Valley near London, although now it is only to found in the north and north-west of Britain. It has been recorded in flower as early as February in Hoy, Orkney, while I have found it still out in July in Ardnamurchan, Argyll. Locally plentiful on moist lime-rich rocks and cliffs, it grows in Brecon, on Cader Idris, and in Snowdonia; and on the limestone cliffs of Ingleborough and Pen-y-ghent it is a breathtaking sight, a mass of brilliant reddish purple dripping over the pale rock. It is one of the delights of Ben Lawers, Clova, Ben Vrackie, and the whole limestone chain from Inchnadamph north to Durness and Bettyhill where it grows with Moss Campion *(Silene acaulis)* and Mountain Avens *(Dryas octopetala)* in blown calcareous sand at sea-level. In Ireland it grows in the Burren, County Clare, the Ben Bulben hills near Sligo, and near the sea in Donegal. The plant is prostrate with small paired leaves which are bluish green in colour, and appear fleshier and softer than the dark-green leaves of the creeping stems of thyme (*Thymus* spp) which it so much resembles and with which it often grows. The flowers are large, 8 to 10 millimetres across, solitary on short stems, and of a rich pinkish-purple colour.

*8-10mm

Cow Parsley

Anthriscus sylvestria (L.) Hoffm.

Identifying umbellifers (plants with umbrella-shaped flower heads) is never simple, but this need not prevent the amateur botanist from tracking down the members of this large and complex family. One of the most useful pointers is the time of year at which each flowers. Cow Parsley is the commonest of the early umbellifers flowering in the southern half of England, where it grows in waste places and road verges. The Orange-tip Butterfly will often rest on Cow Parsley, the green-and-white pattern of the underwing blending superbly with the flowers. Cow Parsley is less common in the Highlands of Scotland and in south-west Ireland. It is a stout perennial with fern-like leaves and hollow, ribbed stems 50 to 125 centimetres high, that are slightly hairy but show none of the purple blotching which is such a feature of the June-flowering Hemlock *(Conium maculatum)*. There are several small bracteoles under each small flower mass but no bracts at the base of the main umbel. The bracteoles of Fool's Parsley *(Aethusa cynapium)*, which flowers in July, are long and drooping. The fruits of umbellifers are most helpful as an aid to identification; those of Cow Parsley are long and smooth with a short beak.

0·5-1·25m

Alexanders

Smyrnium olusatrum L.

Alexanders has a mainly coastal distribution around England, Wales, the Isle of Man, and the easten side of Ireland. North of the Humber it becomes increasingly rare, and it has died out recently in Cumbria. The huge, bushy plants, up to 1·5 metres tall, crowned with spherical umbels of yellow flowers, are a familiar sight on roadside banks and in waste places near the sea, especially where the soil is chalky. Alexanders was introduced as a pot herb, first recorded in 1562 on Steep Holm in the Bristol Channel, and mentioned as Black Potherb in Gerarde's *Herball* of 1597. The young stems are said to taste like celery. The massive stems are solid and strongly furrowed. The leaves are divided into three main lobes, each broadly toothed, dark green, and glossy. The bases of the upper leaf stalks are expanded into broad sheaths which clasp the stem and are boldly and attractively striped. The yellow flower masses which go to make up the umbel are nearly spherical; the fruits are globular, sharply ribbed, and black, a characteristic which led to the name Black Potherb.

0·3-1.5m

Cowslip

Primula veris L.

Cowslip time in April carries with it the feeling that summer is not far away. Sadly, the sight of a field full of yellow Cowslips has become a rarity, not only because of the ploughing and reseeding of old pastures, but also because of the selfishness of those who pick them and dig them up. They are widespread on hedge banks, beside railways, and in grassy places, especially on chalk, limestone, and clay, as far north as the Scottish borders, and then in scattered localities in the east of Scotland, Orkney, Skye, and across the middle of Ireland. In some areas they are still called 'Paigles' or 'St Peter's Keys', a name derived from the old word 'paggle' – to hang down in a bunch. In Switzerland I have heard them called 'Heaven's Keys'. The leaves form a rosette like those of the Primrose, but they are smaller and broader, tapering abruptly to the leaf stalk. The leafless flower stem, up to 30 centimetres tall, may have as many as thirty sweetly scented flowers hanging in a one-sided bunch. Early-purple Orchids are often to be found flowering with Cowslips and, in southern England, they may even be accompanied by the rare Early Spider-orchid *(Ophrys sphegodes)*.

70-300mm

Oxlip

Primula elatior (L.) Hill

Oxlips are the aristocrats among our yellow primulas, and a visit to an Oxlip wood compares with a visit to a cathedral. It is a very local plant, growing in moist woods on chalky boulder clay mainly in north Essex, Suffolk, and Cambridgeshire, where it may be locally abundant. It is a tall plant, up to 30 centimetres high, with pale-yellow flowers in a one-sided cluster. Each flower is shallow dish shaped, not flat like a Primrose, or cup shaped as in the Cowslip. The lobes of the green calyx tube each has a prominent dark-green midrib so that it appears striped. The leaves, which form a rosette, differ from the leaves of Primroses in tapering abruptly into a long-winged stalk. Confusion arises with the False Oxlip which is a hybrid between Cowslip and Primrose *(Primula veris* x *vulgaris)*. This hybrid is not uncommon on downland where both parents occur, but never grows in masses. The leaves of the hybrid taper gradually to the base of the leaf stalk and the flowers droop in every direction. They are richer yellow, and a careful examination of the corolla will reveal folds in the throat, which the Oxlip lacks.

80-300mm

Spring Gentian

Gentiana verna L.

There is something thrilling in the intense blue of gentian flowers, although to enjoy them in Britain entails a visit either to the Upper Teesdale National Nature Reserve or to the lunar landscape of the terraced limestone hills of the Burren in Galway and County Clare. At Widdybank Fell in Teesdale an excellent nature trail starts at Cow Green (OS map 91) and well-laid-out paths enable the visitor to enjoy not only the gentians but other delights which include the pink Bird's-eye Primrose *(Primula farinosa)*, Scottish Asphodel *(Tofieldia pusilla)*, and Teesdale Violet *(Viola rupestris)*. On the Burren and the hills just west of Galway Town, the clumps of bright-blue gentians contrast with the creamy white flowers of Mountain Avens *(Dryas octopetala)*. Near the coast the gentian carpet may even spread off the limestone on to the sand-dune sward. The bright-green leaves are small and blunt, forming a rosette from which arises the solitary flower, 2 to 6 centimetres tall. The five rather angular petals are well separated in a star shape of intense blue, the small white scales at the base between the petal lobes forming a white eye. The flowers close up when the sun is not shining, revealing a narrow, winged calyx which is often tinged with brown.

20-60mm
*15-20mm

Early Forget-me-not

Myosotis ramosissima Rochel

It is easy to overlook the tiny plants of Early Forget-me-not, the smallest and neatest of all the forget-me-nots which grow in Britain, and often only 2 centimetres high. It is more common in the south and east of Britain, but it can be found in dry, impoverished soils and drystone walls as far north as Orkney, rarely elsewhere in northern Scotland, and in scattered localities around the coast of Ireland. The spoon-shaped leaves are often blotched with brownish purple, and form a tight rosette. They are covered with bristly hairs especially along the leaf margins. The flowers are bright blue with a white eye, scarcely 2 milimetres across, and, because the corolla tube is shorter than the calyx, they are not easy to see. Their minute perfection deserves careful examination with a hand-lens. The calyx lobes are pointed and covered with hooked hairs, the calyx teeth spreading out as the fruit ripens. At the same time, the fruiting spike lengthens greatly and becomes much longer than the leafy part of the stem. These two features help to distinguish Early Forget-me-not from dwarfed plants of similar species which can be found in the rather dry, barren places it prefers.

20-150mm

Ivy-leaved Toadflax

Cymbalaria muralis Gaertn., Mey. and Scherb.

Although it is not a native plant, Ivy-leaved Toadflax has spread widely since its introduction over 300 years ago and is now common throughout much of Britain except for the north of Scotland and the Hebrides. It is absent from Orkney and from Shetland. It grows exuberantly on old walls and in dry, rocky places, trailing and rooting in every little crevice it can find. I have found it particularly luxuriant on stabilized shingle near the sea, the lilac-coloured flowers contrasting with White Stonecrop *(Sedum album)* and the yellow of Biting Stonecrop *(S. acre)*. The trailing stems, which may be purplish, bear long-stalked, thick, alternate leaves which resemble those of Ivy *(Hedera helix)*. The flowers are solitary, like small lilac antirrhinums, with two lips. The upper lip is divided into two lobes veined with purple while the lower lip is divided into three lobes and bears a spur at the base. The throat of the flower is blocked by a two-lobed swelling, called the palate, which is developed from the lower lip. It is white with a bright-yellow spot. White-flowered forms have occasionally been recorded from places as far apart as Angus and Somerset.

*7-9mm

Germander Speedwell

Veronica chamaedrys L.

100-250mm

Germander Speedwell or Bird's-eye is very common throughout Britain except in the Hebrides and in Shetland. It grows in grassland, hedges, road verges, and in open woodland. The flowers are handsome, 10 to 12 millimetres in diameter, and of a clear bright blue with a white eye which gives the plant its second name. The stem grows along the ground for a short distance before becoming erect, reaching 20 to 30 centimetres in height. There are two prominent lines of long white hairs on opposite sides of the stem; these comprise an important feature to help differentiate Germander Speedwell from its many relatives. The leaves are sessile, oval and deeply toothed, rather hairy, and dull green in colour. The blue flowers are carried in leafless spikes which grow from the leaf axils on opposite sides near the top of the stem. The seed capsule, hidden within the pointed calyx lobes, is shaped like an upside-down heart, and is flat and hairy. The name Germander is derived from the late Greek, Χαμαιδρυς literally 'on-the-ground oak'. This may be a fanciful comparison of the leaves to those of an oak.

Toothwort

Lathraea squamaria L.

80-300mm

Toothwort is a fascinating and uncommon plant which has no leaves nor chlorophyll and is parasitic on the roots of Hazel, Wych Elm, and rarely on Poplar, Alder, and Beech. It grows throughout central and southern England in moist woods and hedgerows, especially on chalk and limestone, as far north as Perth, and rarely in eastern Ireland. It is most likely to be found in Hazel coppices where it can be remarkably persistent. I have found it flowering in two woods in west Kent where it had been recorded over 100 years before. The underground scaly rhizome bears small rootlets which swell where they attach to the roots of the host plant. The white, downy stem, 8 to 30 centimetres high, often curls over at the tip, unlike the rigidly upright stem of the closely related broomrapes. The flowering spikes of Toothwort often grow in clumps around the base of the Hazel bushes, with Moschatel, Bluebells, and Early-purple Orchids. The showy Purple Toothwort *(L. clandestina)* has been introduced to some as with the roots of Willow and Poplar on which it is parasites.

Henbit Dead-nettle

Lamium amplexicaule L.

50-250mm

Henbit Dead-nettle is a weed of cultivated ground, especially on light, dry soils on chalk and sand. It is distributed mainly on the east side of Britain as far north as Sutherland and it is uncommon in Ireland. It is a branched, straggly plant 15 to 30 centimetres tall, with rounded, toothed leaves, the lower leaves long stalked, the upper leaves half clasping the stem. The leaves and stem lack the irritant hairs of the Stinging Nettle so that this and related plants are given the name dead-nettles. The deep-red flowers are carried in whorls up the stem at the level of the leaf axils, the whorls having few flowers and being well spaced. The sharp-toothed calyx is covered with white hairs and the corolla tube is three to four times as long as the calyx so that the flowers are more prominent than those of Red Dead-nettle. The divided lower lip of each flower may be marked with red, but does not have the prominent reddish-purple spots of Red Dead-nettle. Some of the flowers may remain very small, not opening, and partly hidden within the calyx. Such flowers can be self-pollinating within the unopened flower.

Bugle

Ajuga reptans L.

100-300mm

Bugle grows throughout Britain as far west as St Kilda, but it is absent from Shetland. It has been recorded as high as 450 metres on Ben Griam, Sutherland, but is really at its best in damp grassland and in such places as the edges of woodland rides which are sheltered and wet. The plant has a short rhizome and a series of long, leafy stolons which root strongly from the leaf nodes. The leaves are dark green, faintly toothed, and shiny, often with a metallic, bronzy sheen which is most attractive. The stem, 15 to 25 centimetres high, is hairy on two opposite sides. The flowers are pale blue, rarely pink or white, and are carried in whorls around the stem. They are longer than the bracts which are often purple tinged. During May, when Bugle is still in bloom, the flowers have a strong attraction for Pearl-bordered Fritillaries *(Clossiana euphrosyne)*, and the blue spikes are often clustered with these beautiful butterflies. The rare Pyramidal Bugle *(A. pyramidalis)* grows in crevices on basic rocks in the Burren in County Clare and Galway, in the north-west of Scotland, the Hebrides, and one locality in Cumbria. It is a grey, hairy plant with unstalked lower leaves in a rosette and a very leafy spike in which the darker-blue flowers are rather hidden.

Moschatel

Adoxa moschatellina L.

50–120mm

Moschatel grows throughout England and Wales in woods, copses, and hedge banks wherever the ground is fairly rich, moist, and shady. It is never common but, because of its creeping, fleshy rhizomes, it can spread over a large area where it does occur. It is less frequent in north and west Scotland, where it also grows in damp, shady places among rocks up to 1200 metres, and it is absent from Sutherland, Orkney, and Shetland. Moschatel does not occur in the Isle of Man but is known from one small area of Ireland in County Antrim. The head of small green flowers is cube shaped, with a flower on each of the four side faces and another set on top; this remarkable shape has given it the names 'Town Hall Clock' and 'Five-faced Bishop'. The flowers on the sides each have three sepals, five petals, and five deeply cleft stamens, while the terminal flower has two sepals, four petals, and four divided stamens. The fruit is like a drooping green berry. Moschatel can be confused with fruiting Wood Anemone but the leaves of Wood Anemone are darker green and hairy and the plant grows in dryer parts of the woods.

Dandelion

Taraxacum officinale Weber. agg.

150–300mm

The name derives from the French *dent de lion* and refers to the broad, downwardly pointing teeth along the edges of the leaves. The plant has a strong tap root and a rosette of fairly long, toothed leaves which, when young, can be eaten as a salad vegetable. The flower stem is smooth and hollow with a single golden-yellow flower 15 to 35 millimetres in diameter. Dandelions are a familiar sight everywhere on road verges and in waste places in April. The flower head is made up of masses of yellow ray florets which are notched at the tip, and it is surrounded by a collar of green bracts called phyllaries, the outer ones narrow and recurved. The showy flowers appear to be able to set seed sexually in the normal way but can produce perfect seed without fertilization occurring by a process called apomixis. The seeds of such plants will reproduce the parent exactly, so that distinct local races can easily arise. Indeed, 132 microspecies have been described, and distinct forms will be found in marshes, bogs, and on mountains. The separation of these microspecies is difficult, so we have used the aggregate scientific name to cover them all. The seed of the Dandelion is a long-beaked achene topped with a pappus of fine hairs.

Spring Squill

Scilla verna Huds.

Spring Squill is only to be found in short, grassy turf near the sea, especially alongside cliff-top paths on the west and north-west coast of Britain from Cornwall to Caithness. In Orkney and Shetland it is particularly fine, growing in drifts like tiny Bluebells, and it grows in similar sites on many of the Hebridean islands and on the coast of Ireland between Dublin and Derry. In Pembrokeshire Bluebells and Spring Squills grow on the island of Skokholm on short turf among the Puffin *(Fratercula arctica)* warrens. In Sutherland it can be found with the first flowering of Scottish Primrose. The plants are small and stout, with narrow, grass-like leaves growing from a small bulb. The leaves appear first and tend initially to lie flat on the ground and curl round. The flower spike has up to twelve blue flowers, like six-pointed stars, the anthers being an attractive violet-blue. The rare Autumn Squill *(S. autumnalis)* is only likely to be found with Spring Squill on the coast of Cornwall and South Devon but, by the time it flowers in July to September, all the Spring Squills will be in fruit, the seed pods full of hard, shining black seeds.

50-100mm

Bluebell

Hyacinthoides non-scripta (L). Chouard ex Rothm.

A woodland carpeted with Bluebells is one of the finest sights that springtime offers. The blue is of such an extraordinary quality that the woods seem full of blue smoke. Bluebells can be found in damp, shady fields, hedgerows, and woods throughout Britain, flowering in masses where clearings are created. In the west they grow on sea cliffs and on mountains, and I have seen them forming a blue carpet on the hills of the Isle of Man in an area recently cleared of bracken. They are absent from Orkney and Shetland, apart from an abundant and probably introduced colony in the Manse garden at Lerwick! Bluebells grow each spring from a bulb deep in the soil, relying on a rapid rise in surface temperature to make the necessary growth before they are overwhelmed by other, larger plants. They are absent from the highest parts of the north-east Highlands; the deep layers of soil where the bulbs would lie remain frozen for too long. The narrow, shiny, keeled leaves and smooth stem ooze a slimy mucilage if broken or crushed. Bluebell flowers are sweetly scented. There are two blue bracts at the base of each flower stalk. In every colony of moderate size a few white flowers will be found.

200-500mm

Ramsons

Allium ursinum L.

Ramsons grows in damp woods and shady places, especially near water, throughout Britain except for Orkney and Shetland. It is our only woodland garlic and grows best on basic or lime-rich soils. It is less frequent in north-east Scotland and in Ireland. The most noteworthy feature of Ramsons is the pungent smell of garlic which becomes overpowering if leaves or stems are crushed, and it is enough merely to walk by a wood to be aware that Ramsons grows there. The plants are robust with a clump of broad, bright-green, pointed leaves, the triangular leaf stalks being winged and twisted through 180 degrees. The buds are enclosed in two papery bracts called the spathe, which cover the flowers like cupped hands. As many as twenty-five white, starry flower heads may grow from one clump of leaves. Ramsons is a very invasive, dominant plant and often grows in a carpet of white flowers and green leaves to the exclusion of all else. Cattle will eat Ramsons and, although it does not cause illness, it imparts a strong garlic taint to their milk rendering it unfit for use. They can acquire a compulsive liking for it and will break down fences to get back into the woods where it grows.

150-400mm

Early Spider-orchid

Ophrys sphegodes Miller

The Early Spider-orchid is a rare plant of short turf on chalk and limestone in Kent, Sussex, Dorset, Hampshire, and Gloucestershire. It cannot tolerate disturbance and so will only be found in places which have not been cultivated for many years, yet where sheep grazing or other factors have maintained a fine short sward. Three or four grey-green, blunt leaves form a loose rosette from which the single flower stem grows. Plants growing near the sea can be dwarfed, barely 5 centimetres tall, and very easily overlooked, but normally they are taller and carry up to eight large flowers. The flowers have an extraordinary resemblance to a fat, furry brown spider; the body is represented by the rounded lip, or labellum, which is velvety brown marked with an irregular blue-grey H and has furry brown humps on either side. The two short upper petals are yellow edged with brown, while the sepals are longer and clearer yellow. There can be considerable variation in colour and markings between different plants and, when the flowers have passed their prime, they rapidly fade so that one seldom finds all the flowers on a spike fully out at the same time. Under no circumstances should this species be picked or uprooted.

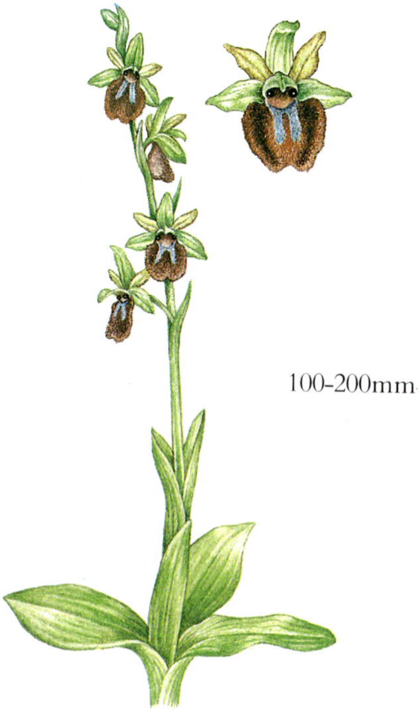

100-200mm

Early-purple Orchid *Orchis mascula* L.

The Early-purple Orchid is one of our commonest orchids, and grows throughout Britain in a wide variety of base-rich sites, ranging from woods and downland to sea cliffs. It can be abundant in some years, but numbers fluctuate widely because it usually dies after flowering and depends upon seed production to provide future generations. The seed takes more than five years to produce a flowering plant. The glossy leaves grow in a rosette and are strongly marked with purple spots, the flower spikes of woodland plants tending to be taller and less dense. The flowers are reddish purple, rarely pink or white, with a stout, upcurved spur, and smell strongly of tomcats. The pollination mechanism is fascinating and can be demonstrated by gently inserting a pencil tip into the throat of the flower so that the sticky disc joined to two pollen masses, or pollinia, grasps the pencil tip. After several minutes the stalks of the pollinia will bend forwards and downwards. A bee visiting the flower will collect the pollinia on its head or mouthparts. By the time it has flown to another plant the pollinia will have swivelled into the low position so that they contact the stigma, thus ensuring cross-fertilization.

150-600mm

Lords-and-ladies *Arum maculatum* L.

Lords-and-ladies is plentiful in England and Wales, growing in ditches, shady hedges, and woods, but it is sparsely distributed in southern Scotland and Ireland, and absent from the Highlands and islands. The large, bright-green, arrow-shaped leaves appear in January well before the flowers. They are often spotted with purple. The flower head is enveloped in a large, pointed, pale-green hood called the spathe, which may also be spotted with purple, and is waisted near its base. The flower head is a specialized structure called the spadix. It is club shaped with a zone of female flowers at the base separated by a zone of barren flowers from an upper zone of male flowers, none of the flowers having any sepals or petals. The upper part of the spadix is barren and purple, rarely yellow. The curious shape of the flower has given it the name 'Jack-in-the-pulpit', among others. The poisonous fruits form an erect cluster of attractive, shiny, orange-red berries which are revealed when the spathe shrivels. *A. italicum* ssp *neglectum* is a rare plant which grows near the south coast. The unspotted leaves appear in November, and the spadix is always yellow.

300-500mm

MAY

The month of May has so much to offer the lover of wild flowers that the difficulty is to know where to go first. On the cliffs by the sea in the west and south-west, Thrift is fully out, with early Sea Campion and Buck's-horn Plantain; the delightful little blue Spring Squill is in full flower in the cliff-top turf, in the south-west and in Orkney and Shetland while, in Sutherland and Orkney, the first crop of Scottish Primroses is in flower, with Roseroot on the cliffs a few metres away. On the downs in the south Burnt Orchids hide among the Milkwort and Salad Burnet and, in favoured spots, Pasqueflowers will be at their best, and Man Orchids on the rougher ridges of the downland slopes.

By the roadsides, Oxeye Daisy, Common Comfrey, and the crane's-bills are all in flower, with Clary on the dry banks on chalky soil, and Deadly Nightshade in the scrub on the woodland edges.

The lowland woods are at their finest, with White and Narrow-leaved Helleborines, Common Twayblade, and, in Kent, the rare Lady Orchid. Herb-Paris, Lily-of-the-valley, and Common Solomon's-seal all flower now, with Angular Solomon's-seal in a few places in the north and west, and the diminutive May Lily, with a spire of white flowers and broad, alternate leaves, flowers in a very few woods in Lincolnshire and Yorkshire.

Globeflowers will be out in the limestone hills of the north, with Mountain Pansy and Bird's-eye Primrose in the high meadows. The curious plant Baneberry should be looked for in the limestone pavement areas, flowering deep down in the cracks between the blocks of stone. On the high cliffs and in the mountains, the saxifrages are coming out; patches of creamy white Mossy Saxifrage grow on the ledges with Roseroot, and the less showy rosettes of Starry Saxifrage on damp ledges where water trickles down the rock face.

On the high tops of Scotland, Trailing Azalea, Bog Bilberry, Bearberry, and Alpine Bearberry with its strikingly veined leaves and reddish stems are in flower now.

In moorland bogs, Bogbean is a mass of pink and white, the purple flowers of Common Butterwort will be found everywhere in the damp sphagnum moss and, in a few of the northern 'mosses' the lovely pink Bog Rosemary is in flower. Bog Rosemary also flowers in Ireland, and this is the best time of year to visit the fabulous Burren country in the west. The Spring Gentians may be nearly over, but this is the season for Hoary Rock-rose, Large-flowered Butterwort, Dense-flowered Orchid, and the many perplexing marsh-orchids with which the area abounds.

Globeflower

Trollius europaeus L.

This tall, stately member of the Buttercup family is immediately recognizable by its pale-yellow, spherical flowers. The true petals are small and narrow, hidden with the many stamens by the ten or so petal-like sepals which curl over them and create the globe shape which gives the flower its name. Globeflower is an upland plant of damp, shady pastures and and wooded gills, such as exist at the foot of eroded cliffs, particularly on limestone and base-rich soils. There it is often accompanied by Wood Crane's-bill *(Geranium sylvaticum)* and Water Avens *(Geum rivale),* an association which is also typical of northern woodlands and meadows in Scandinavia. It grows throughout the centre of Wales, in Yorkshire and Cumbria, and especially up the west side of mainland Scotland, and in north-west Ireland. In Ribblesdale I particularly remember a triangular area, bounded by stone walls, completely filled with a mass of pale-yellow globes. Globeflower grows on sheltered mountain ledges where plenty of humus has accumulated, and it is plentiful at sea-level in a few places in north Sutherland growing with lime-loving plants such as Mountain Avens and Purple Saxifrage.

100-600mm

Pasqueflower

Pulsatilla vulgaris Mill.

I shall never forget my first sight of native Pasqueflower over twenty-five years ago, growing in profusion on the short chalk turf of a rounded hill in the Berkshire Downs. 'Dane's Blood' they were called, the old story being that the purple flowers sprang up wherever the Danes were defeated in their battles with the Saxons. They have suffered much from picking and uprooting, and are restricted and decreasing on unploughed chalk and limestone grassland from the downs of Berkshire, Buckinghamshire, and Wiltshire, the Cotswolds in Gloucestershire, through Northamptonshire to Lincolnshire. The whole plant is silkily hairy, with finely divided, feathery leaves and a collar of hairy divided bracts below the solitary flowers which are fat and erect in bud. The flowers are bell shaped, with a mass of golden stamens in the middle, the purple sepals opening wide in full sunshine and drooping in dull weather. The flower stem elongates in fruit, so that the mass of long, silky-awned achenes stands erect well above the bract collar, which is sometimes tinged with purple. The old name seems to have been Pas or Passe Flower; in 1597 Gerarde subtly altered it to Pasqueflower, observing that it flowered at Easter time.

100-300mm

Meadow Buttercup *Ranunculus acris* L.

Meadow Buttercup is our commonest and tallest buttercup, flowering in damp grassland, on road verges, and in hay meadows throughout Britain, from sea-level to nearly 1300 metres in Scotland. The typical plant is stout, tall, and rather hairy both on the stems and palmate leaves, with a branched flowering stem with several flowers on it. The flowers have five green, elliptical sepals spreading below the longer, round petals. The shiny petals are golden yellow with a nectary covered by a tiny scale at the base of each petal. The fruits are small, smooth achenes, each with a short beak, clustered in a spherical mass on the top of the flower stem. The other two common buttercups, which could be confused with Meadow Buttercup, are Creeping Buttercup *(R. repens)* which prefers damp, muddy disturbed ground, and Bulbous Buttercup *(R. bulbosus)* which grows in dry, lime-rich grassland, such as downland and shell sand dunes. Both have the lower leaves divided into three-lobed segments, Creeping Buttercup spreading by long rooting stolons, Bulbous Buttercup having a swollen stem base and sepals folded downwards.

150-750mm

Greater Celandine *Chelidonium majus* L.

Greater Celandine is not related to Lesser Celandine; it is a member of the Poppy family, Papaveraceae, and looks much like a small edition of the Welsh Poppy *(Meconopsis cambrica)*. It is plentiful in hedgerows, on banks and walls, in waste places throughout lowland England, but scarce in mainland Scotland and Ireland. Although it is native in many places, it has certainly been widely introduced because of its medicinal properties, and will often be found near buildings. The plant is poisonous and will cause purgation and vomiting but an extract, Chelidon Compositum, is still used effectively in the treatment of cystitis. Greater Celandine is sparsely hairy, the yellow-green, pinnate leaves having a greyish bloom on the undersurface. The flowers are borne in clusters on long stems, the two green sepals falling soon after the flower opens. There are four papery yellow petals. The seeds attract ants which carry them away to eat the spongy appendage on the side of the seed, and so distribute them. The sap is a yellow latex, turning red on exposure to air; the Latin name *Chelidonium* likens this red-brown colour to the colour of a Swallow's throat.

230-900mm

Common Fumitory *Fumaria officinalis* L.

100-500mm

Common Fumitory is a widespread plant of arable and waste ground, more common on the eastern side of England and Scotland, particularly on chalky soils. It is not uncommon as a garden weed, and thrives on the edges of cornfields where the rather brittle, scrambling stems are protected by the growing corn. The leaves are finely divided and fern-like, with a bluish tinge and, where it grows extensively, it makes the ground appear hazy – the old name *Fumus Terrae* means 'Earth Smoke'. The flowers are small and tubular, as many as forty on opposite sides of a long spike, the four petals in two partially joined whorls. The top outer petal has a fat spur projecting back beyond the flower stalk and two wings on either side of the petal tip; the lower petal is shaped like a narrow spoon. The tips of the pink petals are blackish red, a feature common to all the fumitories but absent from the similarly shaped genus *Corydalis*. The other species of fumitory which grow in Britain are differentiated by the colour and position of the lateral wings of the upper lip, the shape of the lower lip, and the shape of the fruit. They are uncommon and found mostly in the west and south-west, particularly in Cornwall.

Field Penny-cress *Thlaspi arvense* L.

150-500mm

Field Penny-cress is the most easily identified of the small, white-flowering crucifers, because of the unique shape of the seed capsules. It is most common in central and south-eastern Britain, where it can be a nuisance as a weed of arable land. It can also be found in scattered localities throughout Scotland, Orkney, Skye, the western isles, and Ireland, but it is nowhere common in the north. The lower leaves are undivided and toothed, shrivelling early so that there is no basal rosette such as is common in many crucifers. The upper leaves are arrow shaped and clasp the stem, all parts of the plant having a foetid smell. The small white flowers, with petals lightly notched at the tip, are borne in a long leafless spike. The yellow anthers distinguish it from Alpine Penny-cress *(T. alpestre)* which has violet anthers. The flat seed capsules are very large for such a small plant, up to 22 millimetres in diameter. They are formed by two conjoined cells, each of which has a wide translucent wing, the remnant of the style persisting in a deep cleft between the upper parts of the wings.

Mountain Pansy

Viola lutea Huds.

50-150mm

Mountain Pansy is a true alpine, growing as high as 1150 metres in the Lawers range in the Highlands, with a patchy pattern of distribution northwards from central Wales. It is not uncommon on the limestone of Yorkshire, especially around Malham, in the Pennines and on the Derbyshire limestone, and in many areas through southern and central Scotland except in the far north where it is rare. In Ireland it grows in Carlow, Kildare, and Wicklow, and in the west on the coast of the Burren in County Clare. It is a plant of hill pastures and mountain grassland, occasionally on rock ledges, and is most abundant in areas which are not strongly calcareous, but where the lime has been somewhat leached from the surface soil. The large flowers are most commonly a clear yellow, the three lower petals marked with radiating black lines like a cat's whiskers; they will also be found with the two upper petals mauve and the rest yellow. The variety *amoena* which predominates locally in many northern hill areas has all-mauve flowers, with a yellow base to the lowest petal. Confusing hybrids with Wild Pansy *(V. tricolor)* have been recorded.

Common Milkwort

Polygala vulgaris L.

70-250mm

Common Milkwort grows throughout Britain and Ireland in grassy places on high base content soils, especially on the calcareous soils of the downs of south and south-east England. It is a low-growing plant with no basal rosette and a straggly stem with pointed alternate leaves. The flowers are deep blue although plants with pink or white flowers are not uncommon. They have five sepals, three small outer, and two broad inner ones, often tinged with green, which enfold the petals like a pair of wings. The three petals form a narrow tube with a fringed end which is often white. On peaty, acid soils Common Milkwort is replaced by the equally common Heath Milkwort *(P. serpyllifolia)* which has darker-blue flowers and the lower stem leaves opposite. On the chalk of southern England, Chalk Milkwort *(P. calcarea)* grows with Common Milkwort, from which it can be distinguished by its ice-blue flowers and the lower stem leaves in a false rosette formed by the loss of the lowest leaves. Two very rare milkworts are Dwarf Milkwort *(P. amara),* which grows on the limestone of the Craven Pennines and Teesdale, and Kentish Milkwort *(P. austriaca),* which is known only from a few sites in Kent.

Hoary Rock-rose

Helianthemum canum (L.) Baumg.

Hoary Rock-rose is a rare plant of limestone cliffs and pavements in north Wales from Great Ormes Head to Prestatyn and on the east coast of Anglesey, and in Cumbria and Yorkshire in a number of sites. To see this delightful flower at its best, however, you need to go to the Burren in County Clare where it is abundant. It does not grow where the ground has been cultivated or treated with fertilizer, but is best looked for in the shallow grykes where there is shelter but also adequate light. The plants are perennial, with numerous, woody, prostrate stems and very small leaves which are covered with star-shaped hairs and have no stipules. The flower stems carry clusters of golden-yellow flowers with petals which fold back to expose the prominent cluster of yellow stamens. Common Rock-rose *(H. chamaecistus)* grows widely on basic grassland except in north-west Britain and Cornwall, and is known in Ireland only from one site in Donegal. The yellow flowers are twice as large as those of Hoary Rock-rose and the leaves bear stipules. White Rock-rose *(H. appeninum)* is a very rare plant of limestone turf near the sea in Devon, Somerset, and Glamorgan.

50-120mm

Maiden Pink

Dianthus deltoides L.

Maiden Pink is often grown as a rockery plant, so that it can be difficult to decide if plants found in the wild are escapes from cultivation. The wild form has short blunt leaves with a bluish tinge, which are roughly hairy along their margins and on the underside of the midrib. There are short non-flowering shoots and the flowers are usually solitary on long stems. Around the base of the cylindrical, toothed calyx are one to three scales which have long pointed tips. The flowers are not fragrant and close up in dull weather so that, despite their colour, they are very hard to find. The flowers are delicately marked with a cluster of pale spots at the centre of each fringed petal, and a dark band near the base, which forms a ring with its neighbours. Although the colour is normally rose red, very dark red or white flowers occur naturally. Maiden Pink grows in base-rich grassland, particularly on light sandy soils, often on the edge of golf links near the sea. It is a rare plant with a scattered distribution which includes west Kent, Surrey, East Anglia, Radnorshire, the coast of north Wales between Harlech and the River Conwy, Shropshire and Yorkshire north to Angus and Kincardine on the east coast of Scotland.

100-200mm

Dove's-foot Crane's-bill

Geranium molle L.

Dove's-foot Crane's-bill grows throughout Britain on dry, poor soils, in grassy and waste places, stabilized sand dunes, and in cultivated ground. Although it has been recorded as high as 600 metres, it is less common in the mountainous areas of Wales, northern England, and Scotland, and in Ireland it is commoner in the south and west. It can be distinguished from the stork's-bills *(Erodium* species) by the palmately lobed leaves, circular in outline, and by the untwisted beak which develops on the fruit. The stork's-bills have pinnate leaves and the beak of the fruit twists spirally. Dove's-foot Crane's-bill is rather a floppy plant, with stem and leaves covered in long, soft hairs, and small mauve flowers in pairs on short stalks. The plants growing in grassland by the sea are frequently dwarfed in stature, with shorter leaves and flower stalks, so that the flowers are almost hidden in the turf, much like Sea Stork's-bill *(Erodium maritimum)*. The closely related Round-leaved Crane's-bill *(G. rotundifolium)* is much less hairy, with leaves less deeply lobed and paler flowers with scarcely notched petals. It is a much bigger, bushier plant.

50-150mm

Wood-sorrel

Oxalis acetosella L.

Wood-sorrel is common throughout Britain, except for parts of the eastern counties, Orkney, and Shetland. It is often abundant in ash and beech woods, except on very chalky soils. In the north-west, Wood-sorrel grows with ferns and mosses in sheltered, humid birch woods which occupy north-facing rocky gullies in the hills. It also grows in deep sheltered crevices between boulders, in the entrances to caves and on mountain ledges, up to a considerable altitude, where it is often accompanied by Wood Sage *(Teucrium scorodonia)*. It is wonderful how such a delicate-looking flower as Wood-sorrel can thrive in conditions of lashing rain and high wind. Wood-sorrel spreads by a creeping rhizome, which has clusters of overlapping, fleshy scales below the flower stems. The yellow-green leaves are three lobed, each lobe heart shaped and often purplish on the under surface. The lobes tend to droop, and the leaf resembles that of Shamrock. The cup-shaped flowers are long stalked, with five delicate white petals veined with lilac. The drooping seed capsule explodes when ripe, scattering the seeds for a considerable distance.

50-100mm

Horseshoe Vetch

Hippocrepis comosa L.

Horseshoe Vetch is restricted to short turf and cliffs on chalk and limestone, so that it flourishes on the chalk downs and limestone hills from Kent to Somerset in the south of England, and then northwards to Cumbria and Yorkshire. In Scotland it grows in Kincardine, and in Wales on the Gower and coast of Glamorgan in the south and Great Ormes Head in the north. It is absent from Ireland. It is a straggly, low-growing plant, the pinnate leaves bearing four to six pairs of leaflets with a blunt terminal leaflet. As many as twelve small yellow pea flowers radiate from the top of the long flower stalk. The zig-zag seed pods break up when ripe into horseshoe-shaped pieces, each containing a single seed. Common Bird's-foot-trefoil *(Lotus corniculatus)* has a superficial resemblance to it, but has leaves with three broader trefoil leaflets and two at the leaf base which resemble stipules, flowers often red tinged, and straight seed pods. Horseshoe Vetch is the feed plant of caterpillars of two downland butterflies, the Chalkhill Blue *(Lysandra coridon)* and the kingfisher-blue Adonis Blue *(L. bellargus)*.

50-250mm

Silverweed

Potentilla anserina L.

Silverweed grows in all parts of Britain except for the high land of central Scotland, Wales, and northern England. It is a widespread and common plant of poor, damp grassland, roadsides, and waste places. It also grows in sand dunes and on river gravels, and is particularly luxuriant above the high-tide mark on the shingles of many sheltered beaches in Scotland. There it grows with various oraches (*Atriplex* species), Herb-robert *(Geranium robertianum)*, and the deep-blue flowers of Skullcap *(Scutellaria galericulata)*. Creeping, prostrate runners, rooting at the leaf nodes, radiate from a rosette of narrow, deeply toothed, pinnate leaves which are distinct from those of all other cinquefoils. They are covered with silky hairs, so that the undersides of the leaflets are silvery white. The large yellow flowers are usually long stalked, but the leaves and flowers of Silverweed show considerable variations in different habitats, depending upon the degree of exposure. Sand-dune forms may be extraordinarily hairy on both surfaces of the leaves, with tight rosettes and short-stalked flowers, while plants from damp grassland are stronger, taller, and scarcely hairy.

*30-35mm

Mountain Avens

Dryas octopetala L.

Mountain Avens was an abundant plant during the tundra periods following the ice ages, when base-rich soils, so far unleached, were exposed by the retreating ice. Now it is a rare plant on limestone and basic rocks, preferring drier rock ledges to the wet flushed areas, and growing with other alpines such as Purple Saxifrage, Alpine Cinquefoil *(Potentilla crantzii),* and Northern Bedstraw *(Galium boreale).* Mountain Avens grows on limestone hills in scattered sites in Wales, northern England, and Scotland, particularly on the limestone band which extends from Kishorn in Ross to Bettyhill in Sutherland. It is one of the glories of the Burren in County Clare and Galway and is found on the coast of Donegal and Derry. Where it finds suitable conditions it can form extensive sheets, the strong, woody stems growing 30 centimetres or more in a year. The strongly veined leaves resemble small oak leaves, dark green and glossy on the upper surface, white felted underneath. The flowers are magnificent, solitary on hairy stems, with usually eight creamy white petals and a boss of golden stamens in the centre. The seeds have long feathery awns, initially twisted into a cone, but forming a fluffy head when ripe.

20-80mm

Roseroot

Sedum roseum (L.) Scop.

Roseroot grows on sea cliffs and mountains up to 1300 metres on both calcareous and acid rocks, from Cader Idris and Snowdon in north Wales northwards, especially in north-west Scotland and the western isles. In Ireland it is found mainly on the west coast. The thick, pink, fleshy rootstock smells of roses if cut or bruised; from it grows a number of stems with overlapping fleshy leaves. The greyish-green leaves are toothed at the tip and become strongly tinged with pink and orange later in the summer. The flowers are small, with male and female on different plants, in large, flat-topped, yellow clusters which are pollinated by humblebees. The fruits are erect and reddish in colour, often being mistaken for the flowers. Roseroot grows on narrow ledges and in vertical clefts in the rock with typical flush species such as Yellow Saxifrage *(Saxifraga aizoides),* Mossy Saxifrage *(S. hypnoides),* Mountain Sorrel *(Oxyria digyna),* and, on sea cliffs, with Sea Campion *(Silene maritima).* There it is free from damage by grazing animals and, on the coast of north-west Scotland, Roseroot often grows in dense clumps, filling the dank clefts in the sea cliffs with bright-yellow flowers.

80-300mm

Mossy Saxifrage

Saxifraga hypnoides L.

Like many of our alpines, Mossy Saxifrage is a relict from the ice ages, confined to calcareous rocks and cliffs from mid-Wales to the north of Scotland, where it can be abundant. Although it usually grows high up in the moutains, it occurs at sea-level on the south-east coast of Sutherland. It has an interesting site in the south in Cheddar Gorge in Somerset, where it was recorded in 1726. It prefers rather damp rock ledges and is particularly fine on the limestone cliffs below the millstone grit cap of Pen-y-ghent where, as in many northern sites, it produces cascades of white flowers. Mossy Saxifrage forms dense mats composed of floppy rosettes of mainly three-lobed leaves and many creeping barren shoots which may have small bulbils in the leaf axils. The buds of the creamy flowers are sometimes tinged with pink. In Ireland, Mossy Saxifrage grows on the Burren in County Clare, apart from scattered sites mainly in the west on coastal mountains. The similar but rare Irish Saxifrage *(S. rosacea)* has broader lobes to the leaflets and no axillary bulbils and, like Mossy Saxifrage, shows considerable variability in form. The very rare *S. hartii* is confined to the island of Arranmore in Donegal.

50-200mm

White Bryony

Bryonia dioica Jacq.

White Bryony is the only member of the Cucumber family, Cucurbitaceae, native to Britain. It is found chiefly in south and east England, no further west than Somerset and as far north as Northumberland. It is rare in Wales and absent from Ireland. It is especially common on chalk and greensand, growing in scrub and hedgerows where it climbs by strong tendrils like coiled springs. The leaves are broad and palmate, with a papery texture roughened by small spines. Male and female flowers are borne on different plants, both yellow-green with darker veins, the male flowers in a spray and the female flowers in sessile pairs. The fruit is a squashy round berry which turns colour from green to yellow and finally to red. By the time the fruit is ripe and hanging in glistening red clusters, the leaves have usually shrivelled up. Black Bryony *(Tamus communis)* has dark-green, shiny, heart-shaped leaves and firmer red berries in short-stalked sprays. White Bryony is poisonous to animals and humans, and in France, where it is called 'Devil's Turnip' from the shape of the thick white tuber, it has caused poisoning when women have eaten it in the mistaken idea that it will reduce milk secretion at weaning.

to 1·5m

Sheep's Sorrel

Rumex acetosella L.

50-300mm

Sheep's Sorrel is the smallest member of the Dock family, Polygonaceae, common throughout Britain on poor, leached and acid soils such as moorland and dry heaths. It is often tiny and inconspicuous on the bare, sheep-grazed hills of the north and west, where it only shows up in the turf when the leaves and fruit turn red in the late summer. The narrow leaves are arrow shaped with the lateral lobes pointing sideways and even forwards. Sorrel *(R. acetosa)* may often grow with its smaller relative, even high up on mountains, but tends to like rather lusher conditions. The leaves are broader with down-pointing lobes, and the fruit develops a broad membranous wing around it when ripe, so that the flower spike looks denser and larger. Sheep's Sorrel can also be found on the chalk in areas where the lime has been leached out, growing profusely in areas of recently cleared and burned gorse. If sheep or cattle are forced to eat it because of shortage of grazing it can lead to a condition resembling milk fever. The oxalic acid and soluble oxalates in the leaves cause a reduction in the blood calcium levels, the animals becoming comatose and recumbent.

Trailing Azalea

Loiseleuria procumbens (L.) Desv.

*4-5mm

Trailing Azalea grows on dry stony summits in the north and north-west of the Scottish mainland, usually at heights over 800 metres. It prefers bare ground formed of shattered acid rock, where it grows with Bearberry *(Arctostaphylos uva-ursi),* Alpine Bearberry *(Arctuous alpinus),* Dwarf Willow *(Salix herbacea),* and Mountain Crowberry *(Empetrum hermaphroditum).* It is particularly fine on Lochnagar on the border of Aberdeenshire and Perth but can be found on most of the mica-schist mountains of Perth and Angus, and right up to the north in Sutherland, with single sites in both Orkney and Shetland. It is not easily seen, however; the tiny prostrate plants can be overlooked even when they are in flower. It has tough woody stems and masses of small blunt leaves which resemble Thyme leaves with the edges curled under; the new leaves at the shoot tips are often distinctively tinged purplish brown. The flowers have five petals fused together in a typical rhododendron shape only 4 millimetres across. In such a hostile environment as the bleak summits it is not surprising that Trailing Azalea grows slowly, and a stem 5 millimetres in diameter may be over fifty years old.

Bearberry

Arctostaphylos uva-ursi (L.) Spreng.

Bearberry grows in rocky places and on moors in a few sites in northern England and is widespread in Scotland north of a line between Perth and Glasgow. In Ireland it is a common plant on the Burren in County Clare and Galway, and grows in a few places on the north and west coast. On the mountains it is a plant of the heather zone below the summits, growing with Bilberry *(Vaccinium myrtillus)* and Crowberry *(Empetrum nigrum),* and it is quick to colonize the broken edges of road and railway cuttings through rock. It is a strong-growing plant, woody and prostrate, with long trailing stems and leathery untoothed leaves, some of which turn red in the autumn although the plant remains evergreen. The pale-pink flowers grow in clusters at the end of the stems and are strongly scented, attracting bees (*Bombus* species) and a moth *(Anarta cordigera)* which serve to pollinate the flowers. Alpine Bearberry is a much rarer flower of the extreme north-west of Scotland, and in Orkney and Shetland. It grows on bare mountain tops and has beautifully veined and toothed leaves which turn a brilliant red in autumn before they are shed. It has white flowers and the smaller berries turn black when ripe.

Thrift

Armeria maritima (Mill.) Willd.

Thrift grows abundantly on sea cliffs and salt marsh all round the coast of Britain, and also high up on mountains, reaching 1000 metres on many summits in Scotland, Wales, and Ireland. In these places it is often accompanied by 'maritime' plants such as Sea Campion and Sea Plantain *(Plantago maritima).* When the ice retreated after the last ice age, cushion-forming plants such as Thrift were squeezed out by the overpowering broad-leaved trees, and found refuge in rocky areas free from their competitors, places such as sea cliffs and mountain tops. Thrift grows best on the western seaboard of Britain and Ireland, where it carpets the cliff tops with pink flowers and forms a familiar background to the seabird colonies on Orkney, Shetland, and the islands off the Cornish and Pembrokeshire coasts. The plants have a tough woody stock and fragrant flower heads carried on simple stalks above a cushion of narrow leaves. A continental subspecies. *A. maritima* ssp *elongata,* grows in calcareous pastures near Ancaster and Wilsford in Lincolnshire. It has smaller, purple flowers on longer, crooked stems.

Bird's-eye Primrose

Primula farinosa L.

Although Bird's-eye Primrose grows abundantly in the Alps up to 2500 metres, it is not truly alpine in Britain because it is to be found at modest heights on the limestone grassland of Wharfedale, the Craven Pennines, Teesdale, and Cumbria. There it grows in the high pastures, often among tumbled limestone blocks, favouring sheltered depressions and the sides of moist runnels where it can be abundant enough to put a pink flush on the slopes. I remember a small damp quarry so carpeted with lilac-pink blooms that it was difficult to move without treading on them. Bird's-eye Primrose is small but robust, with a tight rosette of toothed leaves which are white and mealy on the underside. The stem likewise is mealy, (the scientific species name, *farinosa,* describes this appearance of being dusted with flour), and carries a flat-topped cluster of clear lilac-pink flowers, each with a bright-yellow eye. Pin-eyed and thrum-eyed flowers exist, as they do in the common Primrose. The smaller, darker-flowered Scottish Primrose grows only on the north coast of Sutherland and Caithness, and in Orkney, so that there is no possibility of confusion with Bird's-eye Primrose.

80–150mm

Water-violet

Hottonia palustris L.

Water-violet is an uncommon and most attractive plant of marshy ditches and pools, where it can grow extensively. It occurs with greatest frequency on the east side of Britain from east Yorkshire south to Sussex, where it flowers in ditches on the Pevensey Levels. It is found in similar sites on the Somerset Levels south of Mendip, and then in a few scattered sites in south and north Wales, and the Lancashire coast. It is rare in Scotland. The mechanical clearing of drainage ditches and the contamination of the water with herbicides and excessive wash-off of fertilizers have led to the disappearance of Water-violet from many of its old haunts. Water-violet has been named for the mauve colour of its flowers, for it is not related to the violets but is a member of the Primrose family, Primulaceae. The leaves are all submerged, with such finely divided pinnate lobes that the leaf outline resembles the skeleton of a flatfish. There is an impressive leafless spire of flowers, made up of a series of spaced whorls of three to eight flowers, each resembling a fragile mauve primrose with a yellow throat. The seed capsules are globular and hang down all round the stem in clusters on long stalks.

150–400mm

Bogbean

Menyanthes trifoliata L.

Bogbean is widespread throughout Britain, especially in the west and north, and in Ireland. It is a plant of shallow water, loch margins, swamps, and acid bogs. In lakes it will be found on the edges only, because it cannot grow in deep water and prefers still water and a rich muddy bottom in which the rhizomes can creep and root. It also grows in pool and wet boggy areas up to 1000 metres on mountains, favouring the wetter central areas beyond the marginal growth of Bog Myrtle *(Myrica gale)* and sphagnum moss. To the bog-trotting botanist, the sight of a mass of the large, shining, three-lobed leaves of Bogbean should be a warning that a smelly immersion is but a short step away, because Bogbean likes free water in which to grow and will not support even a modest human weight. The mud of an acid bog is glutinous and malodorous and once sampled will never be forgotten. Bogbean has a stout, creeping rhizome and, in suitable wet areas, it can form a carpet of leaves, each leaf held well above the water surface by a long sheathing stalk. The flowers grow in dense spikes, each flower opening out from a pink bud into a feathery white star, densely bearded on the inner surface with shaggy white hairs.

100-300mm

Common Comfrey

Symphytum officinale L.

Common Comfrey is a familiar sight by rivers and streams and in damp roadside ditches throughout qngland, particularly in the south. It is a tall and robust plant, growing in big bushy clumps, with broad, roughly hairy, pointed leaves which extend down the stem as wings. The flower head is usually forked and, in bud, is curled round like a bishop's crozier. The colour of the bell-shaped flowers can vary from white to purple, the purple form appearing commoner in drier localities. Common Comfrey is far less common in the north-west of Scotland and is infrequent in Ireland. In south and east Scotland the most widespread comfrey is Tuberous Comfrey *(S. tuberosum)*; the plant usually has a single unwinged stem and yellowish-white flowers. The hybrid Russian Comfrey *(S. uplandicum = S. asperum* x *officinale)* was introduced as a fodder plant and is now probably more widespread than Common Comfrey. The shape and length of the calyx teeth is one of the most useful features in sorting out the species of comfreys and their hybrids. Infusions of Comfrey leaves have been used for treating lung conditions, and the roots crushed to prepare poultices for curing bruises.

0·3-1·0m

Changing Forget-me-not

Myosotis discolor Pers.

70-200mm

Changing Forget-me-not is a plant of dry open places, growing on heaths, dune grassland, dry rocky hills, and even on walls. It is more common in Scotland, the Hebrides, Orkney, and Shetland, growing abundantly on the coastal grassland of Sutherland and Caithness. It is not so frequent in the Midlands and Wales, nor in Ireland. This delightful forget-me-not is sometimes barely 4 centimetres high and it can be immediately identified by the tiny, short-stalked, yellow flowers, which turn blue-grey as they mature. The flower spike is usually coiled round like a crook at the top, with the mature blue flowers at the base, the yellow flowers and the unopened buds in the coil of the crook. The corolla tube is twice as long as the calyx from which it projects, unlike the short corolla tube of the equally small Early Forget-me-not *(M. ramosissima),* which has flowers that are always blue. I always associate Changing Forget-me-not with the Green-veined Orchid *(Orchis morio),* especially where the latter forsakes the damp clay pastures it so often inhabits and grows instead in stabilized sand dunes near the sea, as it does on the coast of north-west Wales.

Deadly Nightshade

Atropa belladonna L.

0·4-1·5m

Deadly Nightshade is a rare plant, native only on chalk and limestone soils, where it grows on the edges of woods and in scrub. It is most common on the downs of Kent and Sussex; elsewhere it grows in scattered sites throughout Britain and very rarely in Ireland. In many places it was probably introduced because, despite its extremely poisonous nature, it was grown for medicinal purposes. All parts of the plant are toxic, especially the roots and seeds, the chief alkaloid involved being hyoscyamine, with small quantities of atropine. Symptoms of poisoning are dryness of the mouth, blurred vision with dilated pupils, and coma leading to death. It is said that rabbits can eat large quantities of Deadly Nightshade leaves without ill effect but their flesh is then poisonous to any animal eating them! Deadly Nightshade is large and bushy, with broad, pointed leaves either alternate up the stem or in pairs comprising a very large and a small leaf on one side of the stem. The drooping, bell-shaped flowers are solitary or in pairs, carried on short stalks in the leaf axils. They are a dingy purple colour, the petals falling to reveal a shining berry like a flat, black cherry set in a green star composed of the five persistent calyx lobes.

Yellow Rattle

Rhinanthus minor L.

Yellow Rattle grows throughout Britain in grassland and by roadsides. It is especially common on chalk, growing robustly on rich soils, but having a dwarfed form in dry sandy soils. The smaller northern and mountain forms are usually self-pollinated, which has led to the development of distinctive local populations, some of which have been given subspecific rank. These forms are found particularly on the hills and mountains of Scotland. Yellow Rattle is semiparasitic on grasses and other herbs although it is quite capable of photosynthesizing some of its own food. The yellow flowers have a hooded upper lip which bears two small, violet-tipped teeth about 1 millimetre long and a lower lip divided into three lobes. The veined green calyx is flattened from side to side, but inflates when the fruit is ripe; the seeds are shed into the dry calyx, which rattles when shaken. Greater Yellow Rattle *(R. serotinus)* is larger and more branched, with bigger teeth more than 2 millimetres long on the upper lip. It is a rare plant and grows in sandy grassland and cornfields from Surrey north to Ross-shire.

80–450mm

Common Butterwort

Pinguicula vulgaris L.

Common Butterwort is a plant of wet heaths, bogs, and mountain flushes throughout the whole of the Highlands of Scotland, the western isles, Orkney and Shetland. It is widespread in most of northern England, north and central Wales, and in the west and north-west of Ireland. Elsewhere it has never been common and the few localities where it grows in the southern half of England are decreasing, as marshes are cleared and drained. It favours areas which are really wet, containing plenty of sphagnum moss, and such plants as Round-leaved Sundew *(Drosera rotundifolia)*. The pale-green leaves with inrolled margins grow in a rosette; the leaf surface is covered with sticky glands which trap small insects. These are then digested by the plant, providing it with a vital source of nitrogen in an environment which is otherwise poor in nitrogenous material. Common Butterwort is usually self-pollinated, in spite of the presence of a long spur apparently adapted for insect pollination. The rarer Pale Butterwort *(P. lusitanica)* has translucent, greyish leaves, and small flowers with short, down-turned spurs. It blooms from early July onwards in south-west England, western Scotland, and western Ireland.

50–150mm

Wild Clary

Salvia horminoides Pourr.

Wild Clary grows chiefly in eastern and southern England on calcareous grassland and dry, chalky banks, appearing sometimes in quantity on newly created surfaces, such as the sides of motorway cuttings through the chalk, with Chicory *(Cichorium intybus)* and Common Bird's-foot-trefoil *(Lotus corniculatus)*. Elsewhere it is a rare casual. It is a hairy, slightly aromatic plant, with wrinkled, jaggedly toothed leaves which always look grey and dusty. The flowers are carried in whorls on a tall stem which is not much branched, the bluish-violet corolla being scarcely twice as long as the calyx, which is covered with long white hairs. Some of the flowers have a very short corolla tube, so that they remain hidden within the calyx. Such flowers may be cleistogamous, being self-pollinated and setting seed without opening. The closely related Meadow Clary *(S. pratensis)* is a very rare plant of calcareous grassland in southern England between Kent and the Cotswolds. It has glorious big blue flowers which open far wider than those of Wild Clary, and have a prominent, arched upper lip; the calyx does not have the long white hairs typical of Wild Clary.

300-800mm

Oxeye Daisy

Leucanthemum vulgare Lam.

Oxeye Daisy is a common flower, growing throughout Britain by roadsides and in grassy waste places, especially where the soil is calcareous. The attractive nature of these large white daisy flowers is reflected in the many local names given them, such as 'Moon-Daisy' and 'Marguerite'. The leaves which form the basal rosette are spoon shaped and long stalked and, like the narrower stem leaves, they are all dark green and toothed. The flower stems are tall and each carries a single large flower. The receptacle in the centre of the flower is dimpled, surrounded by a zone of tubular yellow disc florets, and the numerous white ray florets, each toothed at the tip, make a typical, flat daisy flower up to 5 centimetres in diameter. In downland grass the tall Oxeye Daisies often grow with pink-flushed Dropwort *(Filipendula vulgaris)* and red Ragged-Robin *(Lychnis flos-cuculi)*, making a delightful mosaic of colour. The flowers of Oxeye Daisy seem to attract tiny thrip-like insects, which emerge in hoards from flowers picked and taken into the warmth of a house. Shasta Daisy *(L. maximum)* is a garden plant with a similar, but much larger, flower. It establishes itself in large clumps, especially in such places as railway embankments.

200-700mm

Lily-of-the-valley *Convallaria majalis* L.

Lily-of-the-valley grows in rather dry, deciduous woods on limestone or sandy soils, where it can form extensive patches. It is widely grown as a garden plant for the fragrance of its flowers, but it is native in many parts of England as far north as the borders, and very rarely in central Scotland. In many areas it has decreased due to picking and woodland clearance; as late as 1910 it grew in woods at Bexley in Kent, but has now gone beneath the urban sprawl of south London, a story which can be repeated for many sites in southern England. It is still plentiful in woods in the north of England, especially in such areas as the Dales of west Yorkshire. Lily-of-the-valley can be a shy flowerer, and the leaves must be distinguished from the garlic-smelling leaves of Ramsons. A pink-flowered form was recorded in Somerset from 1872 to 1924, but has now gone. The fruit is a soft, orange-red berry, drooping on a short stalk; very few berries are produced, and most plants spread efficiently by vegetative multiplication of the rhizomes. All parts of the plant are poisonous, and chickens have been killed by eating the berries.

100-200mm

Fritillary *Fritillaria meleagris* L.

A Fritillary meadow in flower is one of the loveliest sights that any botanist can be privileged to see. Fritillaries grow in water meadows, which require a very specialized form of management involving controlled winter flooding which is rarely practised nowadays. Although formerly widespread in central and southern England, Fritillaries now grow only in the valley of the Thames and its tributary streams, in the famous Magdalen Meadows site and others near Oxford, and in the valley of the River Deben in Suffolk. Controlled access to all these sites is essential to avoid irreparable damage to the plants and their environment. The plants are slender, with narrow, glaucous leaves clasping the stem which bends over at the top to support the single lantern-shaped flower. The petals are strongly keeled and brownish purple, marked with extraordinary chequered patterns of dark and light squares. In one large colony which I visited more than 20 per cent of the plants had white flowers in which the chequered pattern was absent, although the keel of each petal was marked with a violet line. Flowers of various shades of pink and plants with two flowers on the stem are not uncommon.

150-300mm

Herb-Paris

Paris quadrifolia L.

Herb-Paris is an uncommon plant of moist woodlands, usually on chalk and limestone, where it can be locally abundant. Unless it is in flower it is easily overlooked, because the leaves resemble those of Dog's Mercury with which it often grows. Herb-Paris is absent from Ireland, Devon and Cornwall, and much of Wales. It is fairly common in the limestone woodlands of Yorkshire, Derbyshire, and Cumbria, but is a very rare plant in Scotland. The curious structure of Herb-Paris has resulted in many local names, including 'True Love-knot' and 'One-Berry'. All parts of the plant show an unusual degree of symmetry, most divisions having four members. The four broad, pointed leaves are borne symmetrically in a whorl near the top of the stem, and above them there is a solitary flower with each of its floral divisions laid out like a four-pointed star, one above the other. The green sepals are lined up with the divisions between the stem leaves, and the very narrow green petals are set between them at 45 degrees. Each sepal and petal has a yellow stamen set into its base, and the whole structure is crowned by a blue-black berry, itself divided into four and topped by four purple styles.

150–400mm

White Helleborine

Cephalanthera damasonium (Miller) Druce

White Helleborine grows mainly in beech woods on chalk in the home counties, especially along the North Downs and in the Chilterns, where it can grow in thousands. The large flowers stand out clearly against the colour of the dead beech leaves. Although plants will grow and flower well in areas of deep shade where few other plants flourish, individuals which grow in sunny clearings or on the edges of woods are often more robust. The creamy flowers, oval leaves, and fat ovary with parallel ridges distinguish it from the rarer Narrow-leaved Helleborine *(C. longifolia)* which has white flowers, long narrow leaves, and a twisted ovary. Narrow-leaved Helleborine grows in many areas of southern England and Wales, but also in Scotland and Ireland where White Helleborine is absent. The flowers of White Helleborine never open widely. A close examination of the specialized lower petal, the lip, reveals five dark-yellow ridges on its upper surface. These are often chewed away by insects which find them attractive but, despite visits by bees and other insects, the flowers are usually self-pollinated and set seed very efficiently.

150–600mm

Common Twayblade

Listera ovata (L.) R Brown

Common Twayblade is one of our commoner orchids, growing in every part of Britain except Shetland. It flowers abundantly in moist woodland on base-rich soils, the plants there being tall and dark green, while those growing in the open grassland of the downs or in dunes are more compact and yellow-green in colour. The name comes from the two broad, oval leaves although occasionally plants with a third leaf will be found. The flowers are small and green but attractive in shape, the lip being forked nearly to midway, so that the flower bears a slight resemblance to the much rarer downland Man Orchid *(Aceras anthropophorum)*. The leaves of the Man Orchid are narrow, however, and the lip of the flower is divided into four lobes, the arms and legs of the 'man', which are frequently foxy red in colour. Common Twayblade is pollinated by small flies, ichneumons, and beetles which are attracted to nectar secreted in a groove in the centre of the lip. The pollen mass is released explosively on to the insect's head, frightening it off to another flower. Plants take at least fourteen years to reach maturity from seed, but vegetative reproduction also occurs.

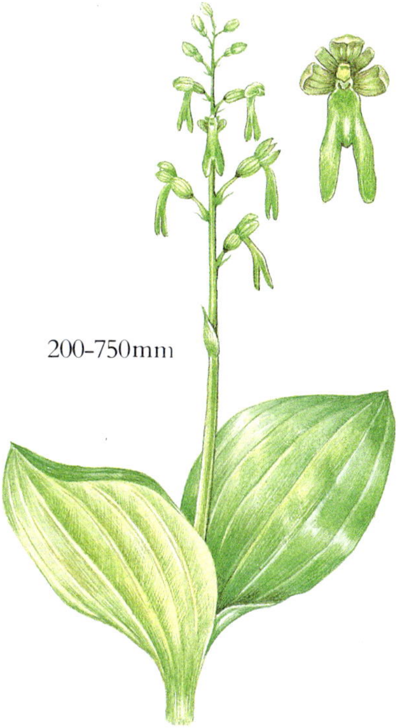

200–750mm

Lady Orchid

Orchis purpurea Huds.

The Lady Orchid is a Kentish speciality, growing in scrubland and beech woods on the chalk in more than ninety sites within the county, from Knockholt in the west to Dover in the east. There have been a very few records of plants flowering in recent years in Surrey, Sussex, and near the Thames in Oxfordshire. It is a remarkably persistent plant, and has been recorded from some sites in Kent for more than 100 years. The Lady Orchid is one of our largest and finest orchids, with a big rosette of broad, shining leaves and a tall spike of as many as fifty flowers. The flower buds have a brownish-red tinge but, when the flowers open out, they are pale pink, rarely white, and shaped like a little lady in a crinoline wearing a poke bonnet. A careful examination of the spots on the lip will show that they are formed by minute tufts of purple hairs. The Lady Orchid flourishes in coppiced hazel woods where it often grows with Early-purple Orchids and Greater Butterfly-orchids *(Platanthera chlorantha)*, but I particularly associate it with the bare terraces below yew trees, growing with Spurge-laurel *(Daphne laureola)* and Gladdon *(Iris foetidissima)*.

0·2–1·0m

Burnt Orchid

Orchis ustulata L.

The Burnt Orchid grows in short turf on chalk downs in the south of England, from Kent west to Gloucestershire and north to Bedfordshire. It also grows on the limestone grassland of north Lincolnshire, Derbyshire, and Cumbria, but is absent from Wales, Scotland, and Ireland. It is like a tiny edition of the Lady Orchid, most plants being less than 7 centimetres tall and, because it takes about sixteen years to reach flowering maturity from seed, it is now only to be found in areas of undisturbed, short calcareous turf which have never been ploughed or fertilized. Such places are old earthworks, barrows, and very steep slopes where it is often accompanied by Chalk Milkwort *(Polygala calcarea)*. It also grows in areas of juniper scrub. The unopened flower buds are rich, red-brown, so that the top of the flower spike appears burned, while the lip of the flower is white with reddish-purple spots. These spots are blotches of colour, and are not formed by clusters of coloured hairs as they are in the Lady Orchid and the Military Orchid *(O. militaris)*. The flowers have a strong sweet scent, and insect pollination results in a high proportion of flowers setting seed.

50-150mm

Green-veined Orchid

Orchis morio L.

Green-veined Orchid gets its name from the dark parallel lines inside the hood of the flower, which are either green or dark bronze in colour. The normal colour of the flowers is dark purple, but they vary from all shades of purple to pink, and even white, the white flowers occurring as less than 1 per cent of any population. In these white individuals the green veins show up particularly well. Green-veined Orchids grow in old damp pastures, in churchyards, and in dune slacks where they can appear in thousands and, unlike Early-purple Orchids, they never grow in woods. They are much smaller, with unspotted leaves and flowers which are sweetly scented and attract insects, especially bees. Green-veined Orchids can be found throughout England as far north as Northumberland, in the coastal area of Wales where they are particularly fine in the dune slacks, and right across the centre of Ireland. Everywhere they are decreasing as old pastures are ploughed and reseeded, so that in recent years it has become a threatened species. The places Green-veined Orchids favour are usually slightly damp, growing with Cowslips and, in many sites, the little Adder's-tongue *(Ophioglossum vulgatum)*.

60-200mm

JUNE

In June the road verges and hedgerows are full of colour, each part of the country having its own particular character. Dry walls and banks along the cliff-top lanes of Pembrokeshire and Cornwall will provide Sea Campion as well as Red Campion, Ragged Robin, cushions of pink-flowered Thrift, and Cliff Spurrey.

In southern England, Honeysuckle and Dog Rose will be coming into flower above verges full of campions, Herb-robert, Fool's Parsley, and the occasional stately and sombre Hemlock. Many of the vetches come out in June, climbing up through the bushes to make a splendid show of mauve and purple flowers. The commonest are Common, Tufted, and Bush Vetch while, in the verge below, one may find the slender, red-flowered Grass Vetchling usually almost hidden in the grass. In much of northern England, the verges are brightened by the showy blue Meadow Crane's-bill, and Downy Rose takes the place of the southern Dog Rose. In upland Scotland, grassy road sides are the places to look for orchids such as Small-white Orchid and both the butterfly-orchids while, in the damp ditches, Early Marsh-orchid and Northern Marsh-orchid can be plentiful.

The seaside is a fascinating place to search for flowers in June. In sandy areas, Burnet Rose, Sea Stork's-bill, Seaside Pansy, Seaside Thistle, and Common Centaury can be found, with Yellow Horned-poppy, Viper's-bugloss, and Biting Stonecrop on shingles. Near the sea in the south the very rare Lizard Orchid flowers on golf links while, at the opposite end of Britain, in Sutherland, the strange Oysterplant flowers on a few shingly beaches. In the north-west, the machair of the Hebrides is a blaze of colour, with Eyebright, White Clover, Bedstraw, Thyme, and the Hebridean Orchid. Yellow Iris is coming into flower in damp areas at the edge of the machair and among rocks behind the beaches.

The southern downs are at their best in June, with carpets of Kidney Vetch and Horse-shoe Vetch, Thyme, Dropwort, Field Scabious, and the charming little Squinancywort. This is the time for the downland orchids, including Fragrant Orchid, Bee and Common Spotted-orchid, and, only in Kent, the magnificent Late Spider-orchid. Weld and Wild Mignonette flower on the track sides, with bright-pink Sainfoin and purple-flowered Hound's-tongue.

In the mountains of the north, every area has its speciality; Moss Campion and Least Willow on bare, leached soils, Alpine Catchfly on limestone, Spring Sandwort and Mountain Everlasting on screes. The high country in June is an endless source of interest.

Lesser Spearwort

Ranunculus flammula L.

The four spearworts which grow in Britain can be distinguished from the other members of the Buttercup family (Ranunculaceae) by their narrow, pointed leaves. Lesser Spearwort is common everywhere in marshes, ditches, and lochs and, because it prefers wet and acid conditions, it is particularly common on the west side of the country. The spear-shaped leaves are slightly toothed near the tips and lie in pairs to one side of the furrowed stem; the flowers are typical small buttercups. In mountainous country a small, prostrate form of Lesser Spearwort grows, with roots forming at some of the nodes on the stem. Creeping Spearwort *(R. reptans)* used to grow by a few gravelly lochs in the north; it was a delicate plant with stems rooting at every node. It seems to have been replaced by the mountain form of Lesser Spearwort, var. *pseudoreptans*. Greater Spearwort *(R. lingua)* is a larger edition of Lesser Spearwort, with a smooth stem and big buttercup flowers. It is an uncommon plant of marshes, ditches, and ponds throughout Britain and central Ireland. Adder's-tongue Spearwort *(R. ophioglossifolius)* is very rare, growing in two marshy sites in Gloucestershire. It has small flowers and spoon-shaped basal leaves.

80-400mm

Common Poppy

Papaver rhoeas L.

Common Poppy grows throughout England in arable fields, roadsides, and wasteland although the use of weedkillers has elimated it from most cornfields. Elsewhere it is less common and, in Ireland, it grows mainly the south. The four poppies which are native to Britain are easily separated by the characteristics of their petals and seed capsules. The scarlet petals of Common Poppy sometimes have a black blotch near the base, and they overlap. The seed capsule is round and smooth; the lid of the capsule is formed by the flattened, rayed stigmas and, when the seeds are ripe, they are shed through small pores below the rim of the stigma disc, scattered like incense from a censer as the capsule blows in the wind. Long-headed Poppy *(P. dubium)* has large, red, overlapping petals with no dark centre, and a long smooth capsule. It is commoner in the north and north-west. Rough Poppy *(P. hybridum)* has small, deep-red flowers, with separated petals, each with a black base, and a bristly, oval capsule. It is a cornfield flower of east and south England. Prickly Poppy *(P. argemone)* has pale, dark-centred flowers and a long bristly capsule. It has a scattered southern distribution.

150-600mm

Weld

Reseda luteola L.

The derivation of the name Weld is uncertain, because it exists in a number of different forms including 'Weild' and 'Would'. The name is mentioned in old writings concerning the dyers, who used Weld to dye cloth yellow and to soften it. For this reason it is also called 'Dyer's Weed' or 'Dyer's Rocket'. It is not uncommon in arable and disturbed waste ground, especially on chalk and limestone, growing in quarries, on cliffs and beside tracks. It is less common in north-west England, northern Scotland, and Ireland. Weld is biennial, forming a rosette of leaves in the first year from which the tall ribbed stem grows. The leaves are narrow and undivided with wavy margins, and the flower spike is stiff and little branched. The short-stalked flowers have four very unequal, divided petals, and a large cluster of up to twenty stamens. Wild Mignonette *(R. lutea)* closely resembles Weld, but has finely divided leaves and longer-stalked, yellower flowers, each with six petals lobed like those of Weld. It is less common, growing mainly in east and south-east England on the open chalk downs or on limestone.

0·3-1·5m

Seaside Pansy

Viola curtisii E. Forst.

Seaside Pansy is a small perennial pansy with a markedly western distribution, growing mainly in stabilized sand dunes among marram grass or on the drier edges of the machair with thyme (*Thymus* species) and eyebright (*Euphrasia* species). Seaside Pansy grows in the dunes in south-west Anglesey, on the coast of Cumbria, and in the dunes on the west coast of Scotland from Sutherland southwards; I have found it in plenty in dunes near Ardnamurchan Point in west Argyll. It is also common in similar habitat in Orkney, in western Ireland, and in Donegal around Lough Neagh. The underground stolons creep through the sand, forming small tufts of rounded, toothed leaves. The lower petals of the flowers are deep golden yellow, while the two upper petals are paler. According to some authorities Seaside Pansy is a subspecies of the Wild Pansy *(V. tricolor)* but it has a spur twice as long, a prominent ovoid capsule in fruit, and in both colour and perennial habit it bears a strong resemblance to the all-yellow form of Mountain Pansy. In Britain these flowers occupy widely separated habitats but in western Ireland they grow together, and the inevitable intermediate forms occur.

30-200mm

Sea Campion

Silene maritima With.

Sea Campion grows on the coast around the whole of Britain most commonly on the west and north-west, in the Hebrides, Orkney, and Shetland, where it is abundant with Thrift on sea cliffs and shingles, and is a feature of the seabird colonies, especially the areas of tumbled rocks and earth where the Fulmars nest. In the south-east of England it is abundant locally, particularly on maritime shingles. Elsewhere it appears in isolated inland sites such as old lead workings on Mendip in Somerset, and in many mountain sites in Wales, the north of England, and the Highlands of Scotland up to heights of 700 metres. Sea Campion is prettier and neater growing than Bladder Campion *(S. vulgaris)*. The two are very closely related, and may represent two subspecies of a single species, but ecologically they are well separated, with Bladder Campion growing on road verges and in waste places inland. Sea Campion forms neat cushions of non-flowering shoots, with small, glaucous leaves of a waxy texture, the large flowers usually growing solitarily and erect on short stalks. The inflated calyx is cylindrical, and does not form the big, round balloon which is typical of Bladder Campion.

80-250mm

Spring Sandwort

Minuartia verna (L.) Hiern.

The bare, desolate spoil heaps of old lead mines are not an obvious attraction for the botanist, but they are particularly favoured by Spring Sandwort, which forms masses of neat cushions made of rosettes of small, pointed leaves, and covered with white, star-like flowers. It also flowers on dry rock ledges and on limestone screes where the ground is well drained. Spring Sandwort grows on the Lizard in Cornwall, among lead workings on Mendip in Somerset, and on the limestone of north Wales. The main area of distribution is in the same type of country in the hills of Yorkshire, Lancashire, Cumbria, and Durham. It is rare in Scotland and, in Ireland, it grows on the Burren in County Clare, in Antrim, and in Derry. Two rare sandworts have similar large, white flowers, but both Arctic Sandwort *(Arenaria norvegica)* and English Sandwort *(A. gothica)* have broad leaves. Confusion can arise between Spring Sandwort and Knotted Pearlwort *(Sagina nodosa)* which also has white, star-like flowers and small, pointed leaves in knobbly clusters up the stem, but it has five styles compared with three in Spring Sandwort.

50-100mm
*8-9mm

Sea Beet

Beta vulgaris L. ssp *maritima* (L.) Arcangeli

Sea Beet is a member of the Goosefoot family (Chenopodiaceae) and is clearly related to Wild Beet, from which a large group of invaluable fodder plants has been developed; these include Spinach, Chard, Mangold, Fodder Beet, and Sugar Beet. Sea Beet is common around the coast of the southern half of England, and in Ireland, but it is rare in Scotland. It grows in a number of different habitats on muddy and shingle shores, on sea walls, and inland beside tidal rivers. It is a tough plant, often semiprostrate where it grows in very exposed sites, and looks very much like garden Spinach gone to seed. The lower leaves are leathery in texture, green and shiny, and, unlike other members of the Goosefoot family, they are never toothed. The stem is tough and ribbed, the stem and leaves frequently becoming tinged with red as the plant matures. The small green flowers are carried in clusters in the axils of the narrow upper stem leaves, the stem being leafy almost to the top, and each flower has five green sepals which curl over at the tips. The sepals thicken up and stick to the fruit, several fruits adhering to form a rough-textured cluster.

0·3-1·0m

Meadow Crane's-bill

Geranium pratense L.

Meadow Crane's-bill is one of the largest and most beautiful of our native crane's-bills, and is locally common as a roadside flower and on the edges of grassy meadows from south-central England northwards. It is absent from Scotland north of the Great Glen and, in Ireland, it is wild only in Antrim. Along roadsides in the Cotswolds and on basic soils in the north of England in Yorkshire, Durham, and Northumberland, Meadow Crane's-bill grows in profusion, the sky-blue flowers making a splash of colour below the grey stone walls. It is a sight denied to those who live in south-east England, where road verges are mowed or sprayed into neat sterility. In a few places in northern England and Scotland it also grows on sea cliffs. The plants of Meadow Crane's-bill are robust, with large leaves deeply divided like the fingers of a hand. The leaves, stem, and calyx are covered in short hairs, and the stipules and tips of the sepals are often tinged with red. The drooping fruits are intriguing; the long beak of the fruit does not twist like the fruits of stork's-bill but each carpel rolls upwards from the base, leaving the seeds attached to the tip like dancers joined by their ribbons to a maypole.

300-800mm

Common Stork's-bill

Erodium cicutarium (L.) L'Hérit.

Common Stork's-bill grows on the coast around most of Britain and Ireland, although it is uncommon in the Hebrides, Orkney, and Shetland. It prefers grassy, sandy places such as sand dunes where it oftens flowers abundantly. In the south and east of England, it flowers inland on sandy commons and heaths, and even on the top of drystone walls. The leaves of Common Stork's-bill are typical of most of the stork's-bills, being pinnately lobed and finely divided; in seaside sand they are frequently rather small and neat, and the whole plant may be prostrate. Leaves and stem are covered in fine hairs which are often dense and sticky on plants found near the sea. The upper petals have a black spot at the base which shows up particularly well in those few plants which have white flowers. The beak of the fruit twists spirally, which is another feature distinguishing the stork's-bills from the straight-beaked crane's-bills. Near the south and west coasts of England and Ireland a sand-dune stork's-bill grows with pale, equal-petalled flowers which have no black spots. It is very hairy and sticky, and has been called *E. glutinosum*. It may be a subspecies of *E. cicutarium* and, where the two grow together, intermediate forms are found.

50–380mm

Ribbed Melilot

Melilotus officinalis (L.) Pall.

The melilots are easily separated from the other members of the Pea family (Leguminosae) by their small flowers borne in long, many flowered spikes. Ribbed Melilot, like all the melilots, is an introduced species, but is now widely naturalized in southern and eastern England on wasteland, roadsides, and grassland. It is particularly common on the chalk where it can be strongly invasive and swamp more delicate species by its exuberant growth. It is less common in the west and north, being rare in Scotland and uncommon in Ireland. The four melilots are distinguished by their flowers and seed pods. Ribbed Melilot has yellow flowers typical of the Pea family, the standard and wings being equal in length and longer than the keel. The pod is brown and markedly wrinkled, giving the plant its name. Tall Melilot *(M. altissima)* has denser, golden-yellow spikes of flowers, with standard, wings, and keel equal, and downy, black pods. Small Melilot *(M. indica)* has tiny, pale-yellow flowers and olive-green pods. White Melilot *(M. alba)* has white flowers with wings and keel shorter than the standard, and brown pods.

0·3–1·2m

Kidney Vetch

Anthyllis vulneraria L.

Kidney Vetch grows in many coastal localities around the whole of Britain, on dry grassland, sand dunes, and on cliffs. It is a common plant especially on the chalk downs of southern England, and grows in mountainous areas on calcareous and base-rich soils. The presence of Kidney Vetch is a firm indication that other interesting plants will be growing nearby. The pinnate leaves have as many as fifteen leaflets, which are often pale and silky underneath, and the terminal leaflet is longer and broader than the others. The round clusters of flowers are often carried in pairs, each with a frill of narrow, green, divided bracts below it. The calyx of each flower is covered in silky white hairs, so that the individual flowers appear to be stuck in cottonwool. The colour is usually yellow but, in plants growing near the sea, flowers can vary from pale cream to deep crimson. The leaves of the seaside plants are often thicker and more fleshy. Kidney Vetch is the food plant for the caterpillars of the tiny Small Blue Butterfly *(Cupido minimus)*. The eggs are laid in the flowerheads, and the caterpillars burrow into the flower. Because of their strong cannibalistic tendencies, only one survives in each flowerhead.

230-300mm

Tufted Vetch

Vicia cracca L.

Tufted Vetch is a common plant, growing throughout Britain except on the high mountains of northern Scotland, in grassy, bushy places, especially in roadside hedges. Unlike the peas (*Lathyrus* species), the vetches have no wings on the stem and the leaves are pinnate with many pairs of leaflets. The leaflets of Tufted Vetch are narrow and, at the tip of the leaf, there is a branched tendril which twists around any supporting object, and helps the plant to climb and scramble. It is a very showy plant, with masses of bluish-purple flowers in long-stalked spikes. The flowers droop and tend to point in the same direction so that the spike appears one-sided, with as many as forty individual flowers on each spike. The seed pods, when they form, are brown and hairless. Bush Vetch *(V. sepium)* is equally common, and grows in much the same sort of habitat. It never climbs to the hedge tops as Tufted Vetch does, and has rather dingy purple, short-stalked flower spikes, each composed of three to eight flowers. The tendrils are branched like those of Tufted Vetch, but the leaflets are broad and blunt, and the seed pods are black.

to 2·0m

Grass Vetchling

Lathyrus nissolia L.

Grass Vetchling is unique among peas flowering in Britain, because the slender, branched plants have no leaflets or tendrils. The narrow, pointed, grass-like 'leaves' are modified midribs called phyllodes and, unless the plant is in flower, it is extremely hard to detect growing in the long grass; for this reason it is probably under-recorded. It is not uncommon in bushy and grassy places such as road verges and railway banks in south-east England and in a few scattered sites between the Bristol Channel in the south-west and north Lincolnshire in the east. There is a single unconfirmed record for the Den of Mains, Angus, in 1849. It favours heavy, base-rich soils such as clay, and I have found it particularly abundant beside marshland roads near the sea in Kent. Seeds of *Lathyrus* species grown abroad are known to be poisonous to horses, causing paralysis of the larynx and hind quarters. Poisoning has been recorded in England in horses fed for ten days on hay containing 85 per cent of seeded Grass Vetchling. They collapsed on exercise but recovered spontaneously.

200-300mm

Dropwort

Filipendula vulgaris Moench

Dropwort is locally abundant on sunny chalk downland and limestone grassland, so that it is common on the downs of Sussex and Kent, and in the west from Dorset and the limestone hills of Somerset, north-east to Yorkshire. Elsewhere it flowers on the coast of Cornwall, in Pembrokeshire, around Great Orme and in Flintshire in north Wales. In the north it is rare, and in Ireland it is known only from the limestone grassland of County Clare. Dropwort is a charming flower, with a rosette of toothed, pinnate leaves, which have as many as twenty pairs of leaflets, with more tiny leaflets in alternate pairs between them. The round flower buds are bright pink, opening out into a delicate spray of creamy, scarcely scented flowers, each with a cluster of long, whiskery stamens at the centre. Dropwort comes into flower at the same time as the best of the downland orchids, which enhances the pleasure of its company. Meadowsweet *(F. ulmaria)* is a plant of wet ditches and marshy places, common throughout Britain. It is a stout and blowsy relative of the downland Dropwort, with a foaming mass of creamy flowers, which have a sickly sweet smell.

100-500mm

Shrubby Cinquefoil

Potentilla fructicosa L.

Shrubby Cinquefoil grew in southern England 10 to 15 000 years ago, when the area was tundra, but it is now restricted to the Lake District, Upper Teesdale, and the Burren in County Clare. In the Lake District it was once known on the screes of Helvellyn, but it is now only to be found on crumbling outcrops of calcareous rock above Wastwater, growing in damp ravines. The locality on river shingles beside the upper Tees is more typical of its European sites; there is grows extensively with Creeping Willow *(Salix repens)* and Melancholy Thistle *(Cirsium heterophyllum)* in bare shingle. In the Burren it is to be found particularly on the edges of the turloughs, the seasonal lakes which are among the fascinating features of the area, and it used to be common in the valley between Ballyvaughan and Lisdoonvarna. Many cultivated varieties of *P. fructicosa* are grown in gardens and admired for their form and colour but the wild plant is just as splendid, with deep-yellow flowers which seem to glow above the grey-green foliage. Even without flowers it is attractive, especially in spring when the clusters of pale leaflets, covered in silvery hairs, break out on the red-brown twigs. Shrubby Cinquefoil usually has a second flowering in August.

0·45–1.0m

Dog Rose

Rosa canina L. agg.

Dog Rose, the unofficial 'English Rose', includes a number of closely related roses; the identification of the numerous species and subspecies of roses which grow in Britain is a job for a specialist. Dog Rose grows in all parts of Britain in woods, hedges, and scrubland, forming bushes on commons, and growing among the rocks on seaside cliffs. The strong stems bear stout, hooked spines and the leaves are smooth, or else hairy only on the underside. The large, scented flowers are white or flushed with pink; the styles at the centre of the numerous yellow stamens are not joined together. The fruits of roses, the hips, are formed by the fleshy receptacle wall, inside which are packed the hard seeds. Rosehip syrup, rich in Vitamin C, can be made from them. Dog Rose hips are smooth and ovoid, the sepals soon falling off from their attachment around the free end of the hip. Among the common roses of the south of England, Field Rose *(R. arvensis)* has the styles joined in a column projecting from the white flowers, and Sweet Briar *(R. rubiginosa)* has pink flowers and leaves smelling of apples when crushed. In northern Britain the commonest rose is Downy Rose *(R. villosa)*, with downy leaves and round, bristly hips.

1-3m

Biting Stonecrop

Sedum acre L.

Biting Stonecrop is widespread and common throughout Britain and Ireland, except for Shetland, especially near the sea, on shingles, in sand dunes and sandy grassland, where it carpets the bare dry ground with bright-yellow flowers. Elsewhere it grows on walls and roofs, and is one of the first flowers to colonize the cracks between the concrete strips on abandoned runways. Biting Stonecrop is perennial, with creeping rooting stems which form dense mats. The leaves are fleshy and stubby, with a bitter, peppery taste, which is indicated by both its common name and the alternative name of 'Wall-pepper'. Tasteless Stonecrop *(S. sexangulare)* has been introduced, and grows on stone walls. The leaves are longer and cylindrical, with no peppery taste; the smaller flowers are paler and first appear in July. Two other yellow-flowered stonecrops both have big clusters of flowers on tall stems, whereas Biting Stonecrop flowers are in groups of two to four. Rock Stonecrop *(S. forsterianum)* is rare on limestone rocks in Wales and south-west England, and Reflexed Stonecrop *(S. reflexum)* is a garden escape, frequently naturalized, with flowerheads drooping in bud, and greyish stem leaves folded downwards.

20-100mm
★10-12mm

Round-leaved Sundew

Drosera rotundifolia L.

Round-leaved Sundew is the commonest of the three sundews, growing throughout Britain in sphagnum bogs and wet, peaty areas. The reddish leaves of Round-leaved Sundew are shaped like spoons on long stalks, and form a flat rosette which is perennial. The upper surface of each leaf is covered with long, glandular hairs, each hair tip glistening with a droplet of sticky fluid. As soon as an insect lands on the leaf, the hairs and then the leaf edge curl round it, to prevent it escaping. The insects are then partially digested with proteolytic enzymes secreted by the leaves, and by this means the plant secures a supply of vital nitrogenous material in an acid environment which is low in soluble nitrates. Oblong-leaved Sundew *(D. intermedia)* is not uncommon in the north and west, more rarely on the Surrey heaths and in the New Forest. It also grows in sphagnum bogs, the short flower spike curving up from below the level of the rosette of narrow leaves. Great Sundew *(D. anglica)* is rare except in the north and west of Scotland and western Ireland, where it grows in wet, gravelly areas as well as bogs. The tall flower spike arises from the centre of the long leaves, which are often erect and shaped like a ski.

50-150mm

Hemlock

Conium maculatum L.

Hemlock occurs in all parts of Britain, but is commoner in the south and south-east of England, where it grows in damp ditches, hedge banks, and disturbed ground. In Scotland it is found more often near the coast, and some confusion can arise from the use of the name to describe other large umbelliferous plants. Hemlock is the only white-flowered umbellifer with a hairless, spotted, hollow stem, which can exceed 2 metres in height. The flowered head is a complex umbel, and beneath the secondary umbels, the upper bracts are borne on the outer side only. The fruits are globular with wavy edges. The leaves and stem give off an unpleasant odour of mice when crushed, and for this reason livestock rarely eat the plant, which is highly toxic. I have known cattle eat Hemlock under drought conditions, with symptoms of drowsiness and paralysis, especially of the hind legs. The alkaloids contained in Hemlock, of which coniine is the most important, are destroyed by slow drying, so that poisoning from Hemlock in hay is unusual. The poisonous quality of Hemlock has been known since ancient times, and it is said to have been the major ingredient of the poisoned cup given to Socrates.

0·6–2·0m

Common Wintergreen

Pyrola minor L.

Five species of wintergreen grow in Britain but, with the exception of Common Wintergreen, they are rare plants of north-east Scotland. Common Wintergreen grows most frequently in open coniferous woodland where there is a good layer of humus and moss. In the south and south-west of England it flowers rarely in deciduous woodland, but in the north it is a plant of pine woods, moors, and mountain ledges up to 1200 metres, provided that there is a reasonable depth of humus. It is rare in Ireland. In the pine woods there may also be Lesser Twayblade *(Listera cordata)* and Creeping Lady's-tresses *(Goodyera repens)*; where it flowers in the dunes of north-east Scotland it is worth looking for the rare Coralroot Orchid *(Corallorhiza trifida)*. The leaves of Common Wintergreen are much paler than those of the other wintergreens but, like them, form a rosette from which the unbranched flower spike arises. The rounded flowers are white or tinged with pink, and the style is short and straight, not protruding from the flower as it does in the other species. The flowers thus have a slight resemblance to Lily-of-the-valley from which it is easily distinguished by the shape of the leaves.

70–300mm

Bog Pimpernel

Anagallis tenella (L.)

Bog Pimpernel is a diminutive and charming plant with creeping, hairless stems, often pink coloured, which root frequently and form mats of growth. It grows in bogs, beside streams, and in damp, peaty places which are often acid, and particularly where there is plenty of moss. The creeping stems seem to burrow through the moss, where the tiny, rounded, paired leaves are virtually hidden so that, until the flowers appear, the plants can be undetected. I have seen Bog Pimpernel produce masses of flowers in the middle of a damp lawn, where it had never been noticed before. Bog Pimpernel has a markedly western distribution, from the New Forest westwards, in Wales, the Lake District, and throughout the west coast of Scotland and the western isles. It is also common in the western half of Ireland. The flowers are borne singly or in pairs on thread-like stalks. They appear to be pink but a close inspection will show that the petals are white, delicately veined with crimson, the flowers opening wide when the sun shines. The globular seed capsule splits round its equator, the top lifting off like the lid of a dish, to reveal the seeds packed inside.

★8–14mm

Common Centaury

Centaurium erythraea Rafn.

Common Centaury grows in dry scrubland, chalk grassland, and sand dunes. It is a common plant across the whole of southern England, flowering less commonly in coastal sand dunes in the north. It is rare in northern and north-eastern Scotland and, in Ireland, it grows mainly in the west. Common Centaury varies greatly in size, from plants 30 centimetres tall growing in sheltered scrub, to tiny, dwarfed plants only a few centimetres high growing in exposed sand dunes on the short turf on cliff tops. About six species of centaury are recognized in Britain, and the identification of the rare species, which all grow near the sea, is a task for an expert. The basal leaves of Common Centaury form a rosette and have three to seven veins. The flowers are virtually stalkless, with a pink corolla tube twice the length of the pointed calyx teeth. Open the flower carefully by slitting the corolla tube from top to bottom; the stamens will be found attached around the top of the tube. Tufted Centaury *(C. captitatum)*, which may only be a variety of *C. erythraea*, has the stamens attached at the bottom of the corolla tube, and Lesser Centaury *(C. pulchellum)* has darker pink, stalked flowers and no basal leaf rosette.

20–400mm

Yellow-wort

Blackstonia perfoliata (L.) Huds.

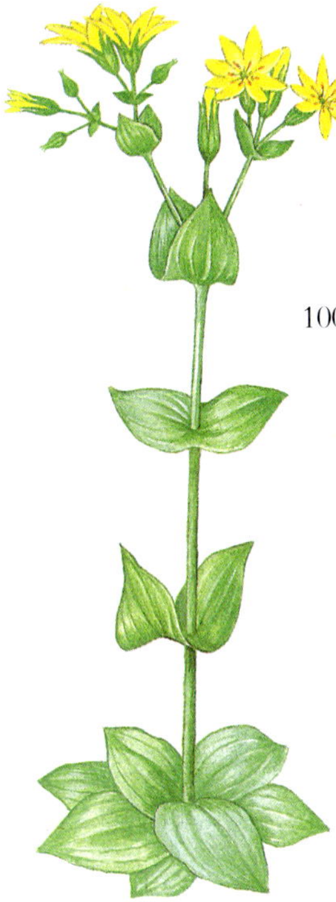

100-400mm

The tall, erect plants of Yellow-wort,, with their bright-yellow, star-like flowers, are one of the delights of the chalk downs of southern England in high summer. The flowers open wide in the sunshine to display the eight petals, which are unique among members of the Gentian family (Gentianacaea) flowering in Britain, four or five petals being the usual number for the gentians and centauries. Yellow-wort grows mainly in grassland and places such a railway banks on chalk and limestone, but also in calcareous sand dunes near the sea. It is a common plant in the whole of south and south-east England, westwards into Pembrokeshire, and in north Wales from Great Orme northwards into Lancashire. Apart from one record in Kirkudbrightshire, Yellow-wort is absent from Scotland and, in Ireland, it flowers on limestone in the mid-west. The leaves of Yellow-wort have a waxy texture which, with their grey-green colour, makes them stand out among the other herbage. The pointed upper stem leaves are joined by their bases to form a collar around the stem, which seems to pierce then, hence the scientific name *'perfoliata'*.

Water Forget-me-not

Myosotis scorpioides L.

150-400mm

Water Forget-me-not is the commonest of four forget-me-nots that grow in or beside water, and has the largest, clear-blue flowers. It can be found throughout Britain in marshes and wet places, by streams, rivers, and beside ponds and lochs. It is less common in the extreme west and north of Britain, and in Ireland. It is a straggly plant with a creeping stem base often immersed in water, and clusters of roots growing out at the nodes. The leaves are long, pointed, and feel smooth, such hairs as they have lying flat on the leaf surface. The spray of flowers has no bracts on it; the tip curls over like a bishop's crozier while the top flowers are still in bud. The triangular calyx teeth are about one-third the length of the calyx tube. The pale-blue petals are shallowly notched at the tip, with white honey guides and a yellow eye to the flower. The roughly hairy Creeping Forget-me-not *(M. secunda)* has leafy bracts at the base of the flower spray, and grows on more acid soils in the west. Tufted Forget-me-not *(M. caespitosa)* grows in chalky areas, and has small flowers and no runners. Pale Forget-me-not *(M. stolonifera)* is rare and northern, with short leaves and small, pale flowers.

Viper's-bugloss

Echium vulgare L.

When flowering en masse on the chalk downs of the south, and on shingles and sand dunes near the sea, Viper's-bugloss is a superb sight, a swathe of vibrant blue. It grows widely in the south of England, and up the east coast as far north as Dornoch Firth in Scotland. In Wales it is not uncommon on the south coast, and then it is found less commonly up the western seaboard to Glasgow, and rarely on the west coast of South Uist and Lewis. In Ireland it is a rare plant of the east coast. In all places it likes a light, well-drained soil in the sun, and flourishes on links and sand dunes, and in disturbed chalkland such as the broken ground of rabbit warrens. The leaves and stout stem are covered in bristles with dark-red or black bases, giving the stem a spotted appearance. The massive flower spike is composed of many curled flower clusters, each producing a series of fine, blue trumpets, so the whole spike remains blue from top to bottom throughout the flowering period. Four of the five pinkish-mauve stamens protrude from the flower, supposedly like a viper's tongue. The name 'bugloss', from the Greek βους γλωδδα, 'cow's tongue',. is a reference to the shape and bristly texture of the leaf.

200-900mm

Bittersweet

Solanum dulcamara L.

Bittersweet is very common throughout England and Wales, except for the most mountainous areas, becoming rare in Scotland and Ireland. It is absent from the Hebrides, Orkney,. and Shetland. Bittersweet is familiar as a scrambling weed among hedges and on the edges of woods, but it also grows on seaside shingles. There the plant is prostrate, without the long, trailing stems of the inland form which die back almost to ground level in the winter. Those on shingle keep a substantial twiggy growth, and deserve more the alternative name of Woody Nightshade. The leaves are slightly downy, the lower leaves resembling those of potato, with a large, pointed main lobe and two smaller basal lobes. The bright-purple, star-like flowers, with a cone of yellow anthers at the centre, are unmistakeable, quite unlike the dingy purple bells of Deadly Nightshade, a name often wrongly applied to Bittersweet. The shiny, egg-shaped berries are first green, but quickly turn yellow and then red. All parts of the plant are poisonous, with a taste which is initially intensely bitter, but leaves a sweet aftertaste in the mouth.

to 2.0m

Rock Speedwell

Veronica fruticans Jacq.

Rock Speedwell is one of the rarest and most delightful of all the speedwells. Finding it will amply repay the considerable physical effort which it will entail. It was first recorded in 1789 by James Dickson, a London nurseryman, on his first plant-hunting trip to Ben Lawers and, to this day, it is only known from that area of Perthshire and from Glen Clova in Angus. Rock Speedwell grows on cliff ledges which run up from the sides of stony, sheltered gullies below the summit of Ben Lawers, and on Meall nan Ptarmachan, where it is accompanied by Alpine Cinquefoil *(Potentilla crantzii),* Alpine Speedwell *(Veronica alpina),* and Alpine Fleabane *(Erigeron borealis).* In Angus it is known from the cliffs of Caenlochan, and from Glen Doll at the head of Glen Clova, where it is again accompanied by Alpine Fleabane, as well as Round-leaved Wintergreen *(Pyrola rotundifolia),* Serrated Wintergreen *(Orthilia secunda),* and a fine display of Mountain Avens. The leaves of Rock Speedwell are small, smooth, and oval; the flowers are cup shaped and deep blue, each petal with a crimson line at the base which gives the flower an attractive red eye.

★10mm

Eyebright

Euphrasia officinalis L. agg.

More than twenty-four microspecies of *Euphrasia* are represented by the aggregate *E. officinalis.* Eyebright is very common as a semiparasite on the roots of grasses, and occurs in a broad variety of habitats, from acid mountain tops, moors and downland, to the short turf of sand dunes and the machair grazings of north-west Scotland and the isles. The variety is amply illustrated by a walk from the top of a mountain such as Ben Hiant in Argyll, to the machair and dunes of nearby Ardnamurchan Point. The montane Eyebrights are tiny, with neat, bronzy leaves and small mauve flowers. Further down the slopes there are plants which are taller and bushier, with bigger, paler flowers. At sea-level and, especially on the summer machair, the plants in more sheltered spots are robust, with large white flowers, veined with purple, and bearing a large yellow spot on the base of the lower lip. Eyebrights are annual, which gives the mountain forms only a short season in which to produce seed. They are usually self-pollinated, so that local variations are more easily produced, and local microspecies tend to arise in isolated mountain communities.

150-230mm

Common Broomrape

Orobanche minor Sm.

Broomrapes are parasitic on the roots of other plants. They have an underground tuber attached to the roots of the host, from which it derives nourishment, and a leafless flower stem bearing a series of pointed bracts. The flower spike in some species is brownish, so that the plant appears dead when in its prime, and the flowers are two lipped, the lower lip being divided into three lobes. Broomrapes are usually named according to the host plant which they habitually parasitize, such as Yarrow Broomrape *(O. purpurea)* or Ivy Broomrape *(O. hederae),* and most are rare. Common Broomrape, however, parasitizes many species, particularly clovers (*Trifolium* species), and will also be found on some members of the Compositae family, such as Smooth Hawk's-beard *(Crepis capillaris)* and Common Cat's-ear *(Hypochoeris radicata).* It is mostly found in southern England. In areas of dense plant growth it can be quite a problem to decide just which plant is acting as host to the broomrape. Two features are useful in separating the various broomrape species; the colour and degree of separation of the two lobes of the stigma, and the position of the hairs on the filaments of the stamens. Common Broomrape has purple, separate stigma lobes, and filaments that are hairy at the base.

100-500mm

Greater Bladderwort

Utricularia vulgaris L. agg.

Bladderworts are floating, aquatic plants which have no root system but have instead a mass of finely divided, submerged leaves with hair-like filaments, and small bladders which are borne on fine stalks in the leaf axils. The bladders serve as floats to keep the plant near the water surface in summer, but fill with water as winter approaches, and sink the plant to the bottom, where it is better protected from the cold. Small water insects become trapped in the bladders, and are digested as a source of nitrogen. The plants also form detached buds, turions, by which they can overwinter. These also serve as a means of vegetative multiplication, because the plants flower sparingly, particularly in the north, and so produce little seed. *U. vulgaris* and *U. australis* are grouped together, because they both have finely divided leaves among which the bladders are carried, and can only be differentiated easily when they flower. Greater Bladderwort is rare and decreasing, in pools, fens, and still lochs throughout Britain. Intermediate Bladderwort *(U. intermedia)* and Lesser Bladderwort *(U. minor)* are also rare, growing in shallow lochs and peaty pools in the north and west.

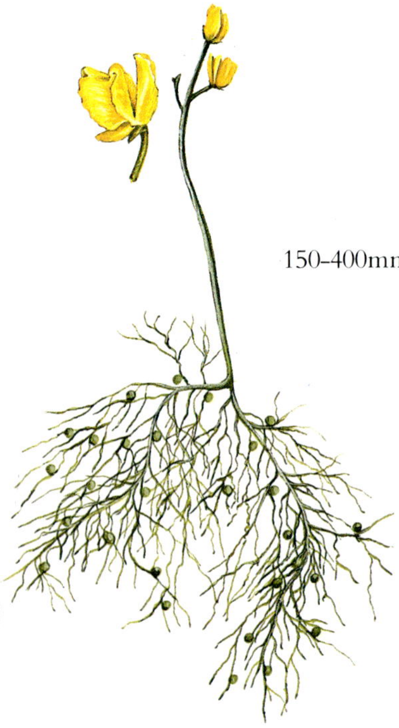

150-400mm

Skullcap

Scutellaria galericulata L.

Skullcap is widespread in south and west England, less so in the north-east and Ireland, and it is absent from Orkney and Shetland. It is a creeping plant, sometimes forming extensive patches, and prefers damp, sheltered places such as stream sides, the margins of lochs, marshes, fens, and wet meadows. Both the skullcaps native to Britain can be recognized by their typical labiate flowers in pairs up the the stem in the axils between the paired, toothed leaves. Lesser Skullcap *(S. minor)* is small, with insignificant, pale-pinkish-purple flowers, and grows more commonly in bogs and wet acid, peaty places. The flowers of Skullcap are larger, dark blue, the lower lip of the flower bearing white patches marked with dark dots. The finest and most intense blue Skullcap I have seen grows with Herb-robert and Wood Sage *(Teucrium scorodonia)* above the high-tide line at the head of the little beaches on the west coast of Argyll. The name Skullcap derives from the shape of the upper lip of the corolla. The older name, Helmet Flower, comes from a fanciful remsemblance of the calyx, when turned upside-down, to a helmet with the visor raised.

100-300mm

Squinancywort

Asperula cynanchica L.

Squinancywort is related to the woodruffs and bedstraws (*Galium* species). It is a common plant of chalk downs, limestone grassland, and calcareous dunes, especially on the downs of Kent and Sussex, westward to Somerset, then across the chalk from the Chilterns to north Norfolk. It is also common on the coast of south Wales, rarely in Denbigh, and in a few sites in Yorkshire and Cumbria. In Ireland it flowers on the limestone of Galway and County Clare, and to the south-west on the Dingle peninsula in Kerry. The sharply angled stem bears whorls of four to six narrow, pointed leaves, two being true leaves and the others leaf-like stipules. The long-stalked flowers are sweetly scented, either white or veined with pink. Although small, they are very pretty, and on the South Downs some plants have flowers so intensely marked with crimson veins that, at a distance, they appear bright pink. Squinancywort is gynodioecious, having female and bisexual flowers on separate plants, and there appears to be some correlation between flower sex and colour. The suffix 'wort' usually refers to a plant with medicinal properties. It was used to treat quinsy, variously spelled 'squinsy' or 'squinancie', a severe throat infection with suppurative tonsillitis.

50-300mm

Twinflower

Linnaea borealis L.

Twinflower was the favourite flower of Carl Linnaeus, the famous eighteenth century Swedish naturalist, and is named after him. It is not uncommon in the forests of northern Europe. In Britain, Twinflower is a rarity, confined to coniferous woodland and shady places among boulders, with Heather *(Calluna vulgaris)* and Bilberry *(Vaccinium myrtillus)*, in north-east Scotland. In Sweden the plants like to trail over mossy covered tree stumps and rocks and, in Glen Clova, Angus, Twinflower flowers in similar sites with Lesser Twayblade. In East Sutherland it grows near Golspie in coniferous woods on sandy soil, with Chickweed Wintergreen *(Trientalis europaea)*, Common Wintergreen and One-flowered Wintergreen *(Moneses uniflora)*, an association typical of the woods where it grows in northern Scandinavia. Twinflower grows in a few other pine wood sites, particularly in Aberdeen and Inverness. The small paired leaves resemble those of Bilberry, with a woody trailing stem which forms extensive mats. The graceful pink flowers hang like paired trumpets from slender pinkish stems, and have a sweet scent.

30-80mm

Field Scabious

Knautia arvensis (L.) Coult.

Field Scabious is common in England and Wales as far north as the Lowlands of Scotland, and then has a scattered distribution in north and west Scotland. In Ireland it is less common, growing in the north-west. Field Scabious grows in dry, grassy places, on chalk downland and calcareous sand dunes. It is the largest of the three blue scabious which grow in Britain, with flowers 3 to 4 centimetres in diameter, on stout, bristly stems. The lower leaves are almost entire, and the upper stem leaves deeply divided. The florets around the edge of the flat flower head are larger than the others, with the two outer corolla lobes much enlarged. The blue stamens, with pink anthers, project from the flowers, dusting everything with red pollen. Small Scabious *(Scabiosa columbaria)* is smaller, less hairy, and has finely divided leaves and pale flowers. The calyx has dark-purple bristles which show through the flower disc. It is common on the chalk downs and dunes, but absent from Ireland, the Isle of Man, and most of Scotland. Devil's-bit Scabious *(Succisa pratensis)* grows in damper ground. It has undivided leaves, sometimes purple blotched, and small, dark-blue, domed flower heads with all the florets of equal size.

0·23–1m

Common Ragwort *Senecio jacobaea* L.

Common Ragwort is a widespread and common weed throughout Britain, by roadsides, in neglected fields and sand dunes, and especially abundant in poor ground. It is usually biennial forming a flat rosette of pinnately divided leaves, with a large terminal lobe to each leaf. The stem is stout, ridged, and rigid, branched mostly at the top of the stem, and the stem leaves are deeply divided, with a small terminal lobe to the leaf. Each flower has a central disc of tubular, hermaphrodite florets, surrounded by about fifteen showy ray florets, which are female. A rayless form, var. *discoides,* occurs occasionally. The seeds, achenes, formed from the disc florets are hairy and those from the ray florets are smooth. Each has a fine-haired pappus which aids seed dispersal by wind, and the fluffy, grey seeding heads, recalling the grey hairs of old men, are the cause of the name '*Senecio*'. Common Ragwort contains several poisonous alkaloids, which cause severe liver damage and death from cirrhosis of the liver weeks or months after animals have eaten it. The leaves are eaten by the black-and-yellow caterpillars of the Cinnabar Moth *(Callimorpha jacobaea),* which often strip the foliage, leaving only the flowers on top of a bare stem.

0·15-1m

Mountain Everlasting *Antennaria dioica* (L.) Gaertn.

Mountain Everlasting is a widespread and typical element of the mountain flora of the north of England, Scotland, and north-west Ireland, particularly in areas of bare, acid rock and shattered summit gravels, where it can be the only attractive plant in an area of botanical dullness. The foliage is neat, with rosettes of grey-green leaves, their edges rolled up to display the white, cottony underside. The dead, shrivelled leaves of previous seasons often form a mat beneath the rosette. Male and female flowers are borne on different plants, and the flowers of one sex often seem to predominate in an area. The male flowers are ringed by white, spreading bracts, which resemble the ray florets of a daisy. The involucral bracts of the female flowers are erect and pink, making the flower head look like a little pink brush. In Shetland, Mountain Everlasting is an important member of the rich cliff-top flora, and elsewhere in Scotland and the Hebrides it grows in sandy links near the sea, and on the machair. It has a scattered distribution in northern England and Wales, and I have found it flourishing in sand dunes on the north coast of Cornwall.

30-200mm

Corn Marigold

Chrysanthemum segetum L.

The widespread use of selective weedkillers in cornfields has resulted in Corn Marigold, like poppies and many other weeds of arable ground, becoming much less common. However, it is widely distributed in Britain on sandy and acid soils, particularly in coastal areas and more especially in the west and north of Britain and in northern Ireland. The finest display of Corn Marigolds nowadays is to be seen in the cornfields and potato fields of the crofts in north-west Scotland and the isles. Although arable farming there can be depressing in terms of yield, it creates a paradise for botanists, with fields yellow with Corn Marigolds and the spectacular pink-and-yellow Large-flowered Hemp-nettle *(Galeopsis speciosa)*. A paradise indeed on a fine June evening, with Corncrakes *(Crex crex)* calling from the nearby hayfields, and Small Tortoiseshell Butterflies *(Aglais urticae)* fluttering around the Marigolds, all against a background of blue sea. The toothed leaves of Corn Marigold are smooth and fleshy, and the bracts beneath the solitary flower heads have broad, pale-brown margins. These features, combined with the splendid yellow flowers 5 centimetres in diameter, make Corn Marigold easy to recognize.

150-500mm

Musk Thistle

Carduus nutans L.

Musk Thistle is an impressive and beautiful thistle, with large, solitary, nodding flowers which are immediately recognizable, and give the plant its scientific name, *'nutans'* – nodding. Musk Thistle is common on chalk and limestone grassland from the downs of Kent and Sussex, west to the limestone grassland of Mendip in Somerset, and northwards to the borders. It is rare in Scotland, flowering in a few sites mainly in the east, as far north as Inverness. In Ireland it is native only in Galway. Elsewhere it is recorded as a rare casual on basic or calcareous soils. Musk Thistle is a large and stout thistle, with prickly wings running down the sides of the stem, but the part of the stem below the flower head is without prickles, and is covered in white cottony hairs. The rows of bract-like structures below the flowers, the phyllaries, form a stiff reddish-purple ruff, each armed with a sharp spine. The outer rows are folded back, giving the nodding flower a particularly graceful appearance. The flowers of Musk Thistle are fragrant, and do not actually smell of musk, but attract many butterflies.

0·3–1·0m

Marsh Thistle

Cirsium palustre (L.) Scop.

Most of our thistles are fairly tall plants, but Marsh Thistle can exceed them all; in warm, sheltered conditions it may grow to over 2 metres in height. Throughout mainland Britain it is a very common plant, growing in marshes, wet grassland, in ditches, in clearings and rides in woods. It flourishes most in damp conditions, commonly on acid soils, being particularly fine in the damp, warm bottoms of stream valleys in Devon and Cornwall, and everywhere it is a great attraction to butterflies, especially fritillaries. It is a little less common on the west coast of the Hebridean islands. Marsh Thistle is stout and erect, somewhat branched towards the top of the stem, with very prickly leaves which are slightly hairy on the upper surface, and have a little cottony down on the underside. The prickly wings on the stem run to the very top beneath the flowers, which are small and egg-shaped, in clusters on short stalks. The phyllaries below the flowers are narrow, often purplish, and tipped with small prickles. They lie closely overlapping each other, so that the flower base looks almost smooth. White-flowered plants are not unusual, and the thistledown of both types is feathery.

0·5-2.0m

Bristly Oxtongue

Picris echioides L.

Precise identification of the many yellow-flowered members of the Compositae family is difficult, but Bristly Oxtongue is one which can be recognized even by the novice. It grows commonly in the southern half of England south of a line from south Wales to Hull on the east coast, and then rarely in north Wales, Lancashire, near Newcastle, and in south-east Scotland. It is a flower of wasteland, roadsides, and field borders, common on the chalk and in the lowlands on clay soils, often near the sea. Bristly Oxtongue is biennial, with a stout, branched, leafy stem and, as the name implies, is well supplied with bristles in every part. The lower leaves are broad and shaped like a cow's tongue, bearing stout bristles which grow from prominent white pimples on the upper surface of the leaf. These bristles are three-pronged, with minute hooks on each point. The stem leaves are toothed and wavy, and similarly bristly. The outer phyllaries around the base of the flower are broad, triangular, and very bristly. When the pale-yellow, Dandelion-like flowers are over, these phyllaries persist like an angular cup around the erect, pointed inner phyllaries, which in turn surround the feathery seed head.

230-700mm

Mouse-ear Hawkweed

Hieracium pilosella L.

The vast genus *Hieracium* contains more than 400 microspecies and, because the mountain forms are self-fertilizing, many of these occur in strictly localized areas. Mouse-ear Hawkweed is found in all parts of Britain except Shetland, growing on grassy banks and walls, on heaths, and in dry, rocky places. In Cumbria I have seen it on the drier mountain ledges, with dwarfed plants of Golden-rod *(Solidago virgaurea),* in the Hebrides growing in sand near the sea, and in the lawn of my own garden in Sussex, where it is all too common. Mouse-ear Hawkweed is one of the least difficult *Hieracium* species to identify. The blunt basal leaves are untoothed and form a rosette, the leaves being dark green above and felted with white hairs underneath. All parts of the plant are liberally covered in long white hairs, and the outside of the phyllaries around the flower bear black-based, glandular hairs as well. Leafy stolons spread above ground from the rosette, and terminate in overwintering rosettes. The pale, lemon-yellow flowers are solitary, and the florets are reddish on the outer surface. A number of varieties of Mouse-ear Hawkweed are recorded.

50-300mm

Water-plantain

Alisma plantago-aquatica L.

Water-plantain is a common plant of ponds, ditches, and muddy places by rivers and streams, wherever the water is still or slow flowing. It grows throughout England, but is less common in the south-west and Wales. In Scotland it is sparsely distributed in the south and east only and, in Ireland, it has scattered distribution in all areas. It is a stately plant with a stout stem devoid of leaves. The leaves which appear above the water surface have long, spongy stalks, and resemble the leaves of Greater Plantain *(Plantago major),* being pale green with a marked pattern of longitudinal veins. The underwater leaves are strap shaped. The flowers are borne in whorls up the stem, each branch divided into smaller whorls. They have three delicate white petals, each flushed with pink at the base. When the fruits ripen, they form a flat-topped ring of closely packed carpels, each with a long style which arises from the base of the carpel on the inside of the ring. Narrow-leaved Water-plantain *(A. lanceolatum)* which is less common and grows in similar sites mainly in southern England, has narrower leaves, smaller, pinker flowers, and short styles which arise near the top of the carpels.

0·3-1·0m

Yellow Iris

Iris pseudacorus L.

0·4–1·5m

Yellow Iris, or Yellow Flag as it is often called, grows in all parts of Britain except in the mountains. It is a common plant, with broad, sword-like leaves and stout creeping rhizomes, and it will grow anywhere which is wet and muddy, by pools and streams, in marshes and damp meadows, often forming extensive patches. It is particularly luxuriant on western coasts. The flower buds, two or three on a stout leafless stem, are enclosed in a broad, papery edged spathe, bursting out from it as they flower, usually in a sequence and not all at the same time. The flowers are typical 'iris' flowers, lacking the beard which is a feature of many of the garden irises. The three broad, yellow outer petals, veined with purple, hang downwards and are called the 'falls'. The three inner petals, the 'standards', are small, erect, and narrow, and the three stigmas are thick and petal-like, the branched tips or 'crests' being longer and more prominent than the inner petals. There is a flap over each stigma point which prevents self-fertilization. The huge, green, three-sided seed capsules are almost as outstanding as the yellow flowers, and remain well into the autumn.

Fragrant Orchid

Gymnadenia conopsea (L.) R. Br.

150–750mm

Fragrant Orchid is one of our commoner orchids, flowering in a variety of different habitats throughout Britain, sometimes appearing in vast numbers. Three different forms exist, which grow in specific types of habitat; the colour also can vary considerably, because albinism is not uncommon. Particularly in the downland from, plants will be found with flowers varying in colour from white, through all shades of lilac and pink, to deep carmine. The normal form growing on downland and limestone grassland has a lower lip with three deeply divided lobes, and is sweetly scented. The nectar can easily be seen through the wall of the long, curved spur, and attracts both bees and moths, the latter being more important as pollinators. In marshes and fens var. *densiflora* is found. It is a large plant with a densely packed spike of flowers, which have a strong scent of carnations. The broad lip has large side lobes and a small, poorly defined, central lobe. The third form, var. *borealis,* is late flowering, and has a markedly northern distribution in hill pastures up to 700 metres, growing even where the soil is slightly acid. The flowers are small, with a tiny lip which is scarcely lobed, and they smell strongly of cloves.

Greater Butterfly-orchid

Platanthera chlorantha (Custer) Reich.

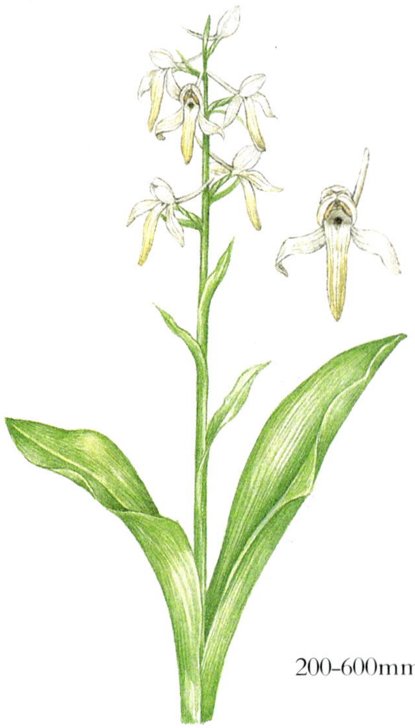

200-600mm

Greater Butterfly-orchid grows not uncommonly in southern England and in the west of Scotland; elsewhere in England, Wales, and Ireland it is far less common. In its southern localities it favours deciduous woodland on base-rich or calcareous soils, especially the damp, mossy woodlands which lie at the foot of the downs. In Wales and Scotland it grows in the permanent hill pastures and beside the roads, where Lesser Butterfly-orchid *(P. bifolia)* also flowers. The two species can be separated by the shape of the throat of the flower, which is the entrance to the spur, and the position of the two pollen-bearing organs, the pollinia. These are club-shaped, and bear at their bases a sticky disc, the viscidium, by which they adhere to the heads of visiting insects. Greater Butterfly-orchid has a wide throat, with pollinia attached to the side and sloping inwards and forwards at the top. The pollinia of Lesser Butterfly-orchid lie parallel and close together, so that the throat of the flower seems closed. In both species the spur is long and full of nectar, and the sweet scent, which is particularly marked at night, attracts night-flying moths, which are highly efficient pollinators.

Bee Orchid

Ophrys apifera Huds.

150-500mm

The Bee Orchid is widely distributed throughout England and Wales, but is absent from much of Scotland. In Ireland it is common on the limestone of the Burren in County Clare, and right across the central plain. It favours short chalk turf and grows well in undisturbed places which are open and sunny, such as old chalkpits and earthworks, dune slacks, and occasionally on clay soils. It is becoming rarer in all its range, because it takes five to eight years to reach flowering maturity from seed, and then may only flower once; those who selfishly pick Bee Orchids or uproot them can be largely blamed for its decline. The pollen-bearing organs, pollinia, are carried in a long structure shaped like a duck's head, which projects above the furry, bee-shaped lip. The pollinia swing down on stalks so that they land on the stigma, and this habitual self-pollination is the main reason for the variants which persist. The so-called Wasp Orchid, *O. apifera* var. *trollii,* has a long pointed lip barred with brown and yellow; var. *flavescens* has white petals and a green lip, and a strange variant, the semipeloric form which is known from one hill in Sussex, has the lip replaced by a structure resembling a pink sepal.

Common Spotted-orchid

Dactylorhiza fuchsii (Druce) Soó

100-450mm

Common Spotted-orchid is truly the commoner of the two spotted-orchids, growing in every part of Britain in many different types of habitat. It grows in vast numbers on chalk downland in the south-east, in chalkpits, light deciduous woods and by roadsides. The degree of leaf spotting can vary from unspotted to dark, dense, purple blotches; the flowers are densely packed into a tapering spike. The lip of the flower is clearly divided into three lobes, the central lobe being triangular, and it is marked by a double loop of broken lines and dots. Two distinct subspecies occur; *D.f.* spp *hebridensis* has been recorded from many of the Hebridean islands, where it flowers profusely on the machair. It is a short, dumpy plant, with large flowers having a deeply lobed lip and heavy markings. When fully out, the flower spike looks like a short, pink drumstick. *D.f.* ssp *okellyi,* which was first found in County Clare in Ireland, is common in north and west Ireland, and has been found in north-west Scotland and Tiree in the Hebrides. It has slender, unspotted leaves and a square-topped spike of white, scented flowers. It is quite distinct from the albino form of Common Spotted-orchid.

Early Marsh-orchid

Dactylorhiza incarnata (L.) Soó

100-300mm

Early Marsh-orchid grows throughout Britain in many different types of damp habitat, ranging from hill pasture, bogs, and marshes, to dune slacks, on alkaline or neutral soil. The erect, keeled leaves are hooded at the tip and are normally unspotted. Five subspecies, which vary greatly in colour, have been recorded, but the flower shape is constant and easy to recognize. The bracts are long and pointed, often marked with spots and lines. The lateral sepals are folded back above the hood formed by the upper sepal and two upper petals. The sides of the lip, which bears a prominent red double loop, are folded back so strongly that the whole flower appears laterally compressed. The normal form is flesh pink in colour. *D.i.* ssp *pulchella* is mauve, commoner on mountains and more acid soils. *D.i.* ssp *coccinea* is a short, stout plant with striking vermilion flowers and grows mainly in dune slacks. *D.i.* ssp *ochroleuca* is a rare creamy yellow form, which has died out from most of its fenland sites, but still flourishes in Surrey. *D.i.* ssp *cruenta* is the only form to have spotted leaves, the spots being concentrated towards the leaf tip. It is known from a wide area around County Clare in Ireland.

JULY

By this time in July, even the high mountains have warmed up; the flower seeker can choose anywhere from sea-level to mountain top, and be sure of finding interesting plants in flower. This is the best time to look for saxifrages, from Yellow Saxifrage which grows commonly in wet flushes and damp roadside ditches throughout much of the Highlands and Lake District, to rarities such as Drooping Saxifrage and Highland Saxifrage which occur on a few Highland peaks only. Alpine Lady's-mantle and Alpine Meadow-rue are easy to find over much of the mountain tops, with Cloudberry on many of the acid uplands, although it seldom bears its delicious berries in any quantity. Golden-rod grows on the mountains in a delightfully neat dwarf form. On the limestone hills, the beautiful Dark-red Helleborine is in flower, often growing out of narrow cracks in the rock face. Bog Asphodel is in full flower now and, in the same boggy areas, Pale Butterwort flowers after its common relative is over, with Grass-of-Parnassus and, in a few sphagnum bogs, the rare and elusive little Bog Orchid.

Marshes and fens are full of flower (and full of biting flies also!) with Water Mint and Meadowsweet, Flowering Rush and Marsh Helleborine. In the fens of Suffolk and Norfolk, Milk Parsley is out, the feed plant of the beautiful Swallowtail Butterfly which is on the wing at this time.

In southern chalk woodlands the first of the late summer helleborines, the Slender-lipped Helleborine, is coming out while, in mossy pine woods in east Scotland and one small area of Norfolk, Creeping Lady's-tresses flowers. On the downland of the south this is the time for thistles to bloom, Dwarf, Carline, and Spear Thistle, with the large purple heads of Greater Knapweed as well. Pyramidal Orchid is the common orchid in flower in the downland grass, with Frog and Musk Orchids less common in very short turf. Road verges on the chalk are graced by the tall mauve spikes of Giant and Nettle-leaved Bellflower, with golden-yellow clumps of Dyer's Greenweed, and the stately mulleins. In arable fields, Common Toadflax, the two fluellens, and Venus's-looking-glass are all in flower, with Large-flowered Hemp-nettle in fields in the north.

The seaside is probably the richest area for flower hunting in July, with Sea Pea and Sea Kale on the shingles in the south, Sea Holly, Portland Spurge, centauries, and many of the strange, parasitic broomrapes on sandy shores and dunes while, in the north, on more rocky shores, Scots Lovage and Bloody Crane's-bill are in full flower.

Sea-kale

Crambe maritima L.

The best places to see Sea-kale are on the south coast shingles, from Suffolk to Cornwall, where it can be locally abundant. It is less common up the west and east coasts as far north as southern Scotland and, in Ireland, it is known only from Wicklow. The plants have a woody root stock which grows from deep down in the shingle where it is moist. Early in the year the first leaves appear. They are crumpled and mauve and, at that stage, are said to be edible but they are tough, rank tasting, and very salty. As the leaves unfurl they lose their mauve colour and become bluish grey; the full-grown leaves are 30 centimetres long, wavy edged, and succulent. In full flower, Sea-kale is a splendid plant; the crowded heads of white, four-petalled flowers are so heavy that they bend to the ground, covering the plant in a cascade of white, slightly scented flowers. In July the shingles are dotted with the white humps of flowering Sea-kale, interspersed with the orange-yellow flowers of Yellow Horned-poppy *(Glaucium flavum)* and, in parts of the east and south-east, the beautiful purple and blue flowers of Sea Pea *(Lathyrus japonicus)*. The fruit of Sea-kale is hard, green, and spherical, containing a single seed.

300-600mm

Perforate St John's-wort

Hypericum perforatum L.

Perforate St John's-wort is common in England and Wales, growing abundantly on the edges of woods, in hedges, grassy places, and on the roadsides, especially on calcareous soils. In southern Scotland it is less common, and elsewhere in Scotland, the Hebrides, Orkney, and Shetland, it is rare or absent. In Ireland it is commoner in the centre of the country. Three important features help to differentiate between the thirteen species of St John's-worts: firstly, the shape of the stem and the presence on it of raised lines; secondly, black dots on the flowers; and thirdly, translucent glands in the leaves – these appear like pinpricks when a leaf is examined against the light. Perforate St John's-wort is the only species to have a round stem with two raised lines on opposite sides. The wavy, yellow petals, twice as long as the pointed sepals, have a few black dots on their margins. The leaves are heavily marked with translucent glands, and may also have a few black dots on them as well. The other, less common, species of St John's-wort can be differentiated by their stems being square or lacking raised lines, by their hairiness, or by the absence of the translucent leaf glands so typical of Perforate St John's-wort.

300-900mm

Common Orache

Atriplex patula L.

100-700mm

The goosefoots and oraches are closely related. Goosefoots have hermaphrodite flowers and a fruit surrounded by a ring of small sepals, while oraches have male and female flowers separate on the same plant and a fruit enclosed in two triangular bracteoles. Common Orache grows in all parts of Britain except on the highest ground, and it is less common in Wales and Scotland. It is a plant of arable and waste ground where the habitat is open and, although it grows by the sea, it has a wider distribution inland than the other oraches which, with the exception of Spear-leaved Orache *(A. hastata),* grow only on the coast. Common Orache is a smooth, much-branched plant, somewhat mealy and sometimes reddish tinged. Near the sea it may become prostrate in growth, and can be confused with the seaside oraches. The triangular leaves are the best means of clear identification, tapering gradually to the leaf stalk, and having two basal lobes which point forwards. This separates them at a glance from the leaves of Spear-leaved Orache, which taper abruptly to the stalk and have lobes which point sideways. The flowers are small and undistinguished, borne on a long, thin, almost leafless stalk.

Bloody Crane's-bill

Geranium sanguineum L.

100-450mm

The showiest member of the large family of geraniums to be found in Britain, Bloody Crane's-bill flowers throughout July and into August. It is widely distributed on dunes, dry calcareous grassland, and on limestone rocks but it is absent from south-east England where any records are likely to prove garden escapes. It makes a vivid splash of colour against the limestone rocks of the Burren in County Clare and Galway, where it is accompanied by Dark-red Helleborine *(Epipactis atrorubens)* and the red-flowered parasite Thyme Broomrape *(Orobanche alba).* On Great Ormes Head, Clwyd, Bloody Crane's-bill and Dark-red Helleborine grow together and, at Kynance Cove on the Lizard, Cornwall, it grows with Thyme Broomrape. It thrives also among the dunes on Anglesey, and Cumbria, and Lancashire, flowering prolifically on dunes on the Northumberland coast with Purple Milk-vetch *(Astragalus danicus),* Lesser Meadow-rue *(Thalictrum minus)* and Scottish Gentian *(Gentianella septentrionalis),* and in many sites scattered up the east coast of Scotland. I have found it in plenty among sand dunes on Coll in the Hebrides, and on cliffs near Ardnamurchan Point in western Argyll.

Common Restharrow

Ononis repens L.

Common Restharrow is a common plant of rough calcareous grassland, roadsides, and sandy places near the sea. It grows around the coast of England, Wales, and eastern Scotland, and inland in the south and east of England. It is rare elsewhere and grows sparsely in the east of Ireland. It spreads by creeping stolons, forming dense patches, the tough roots impeding cultivation and giving rise to its name. It is a hairy plant, with a faintly sticky feel from the glandular hairs on the serrated, three-lobed leaves. The stems are hairy all round, and seldom have spines except in the form var. *horrida* which is usually found near the sea. The wings of the pink, pea-like flowers are longer than the keel, and the little hairy pods are shorter than the calyx. On the island of Alderney in the Channel Islands, Common Restharrow grows in masses, with intensely pink, strongly scented flowers in contrast to the rather dusty, dowdy appearance it has on the chalk downs in southern England. The flowers attract butterflies, especially the Common Blue *(Polyommatus icarus)* and its caterpillars can use Restharrow as a feed plant. Spiny Restharrow *(O. spinosa)* grows mainly in south-eastern England on heavier clay soils, and is absent from Ireland.

*10-15mm
to 300mm

Sea Pea

Lathyrus japonicus Willd.

Sea Pea is restricted almost entirely to shingle beaches where it forms dense patches of dark-green foliage with masses of large purplish flowers, which turn blue as they fade. Although shingle would appear to be a very hostile environment, it supports a fascinating array of summer flowers which are often brightly coloured, as if to make amends for the bleakness of the rest of the year. Sea Pea rarely exceeds 15 centimetres in height, while other plants found with it are also adapted to their dry, exposed situation, and either grow flat on the ground, like the prostrate forms of Bittersweet and Blackthorn *(Prunus spinosa)*, or are mat-forming like Sea Campion. Sea Pea grows on the coast of Sussex where it is one of the many attractions of the Rye Harbour National Nature Reserve, and from Felixstowe to Lowestoft in Suffolk. Elsewhere this handsome pea is rare, occurring in a few widely scattered localities. It is commonly associated with the big cabbage-like plants of Sea-kale, the foetid little Sticky Groundsel, *(Senecio viscosus)*, and Viper's-bugloss although the last is absent from Shetland and Ireland where Sea Pea grows.

230-300mm

Alpine Lady's-mantle

Alchemilla alpina L.

50-230mm

The genus *Alchemilla* contains species which are difficult to separate, but Alpine Lady's-mantle is easy to identify. It grows in mountain pastures, on rock ledges, and on screes up to 1300 metres throughout the Highlands of north-west Scotland, in many mountainous areas of the western isles, and on Ronas Hill in Shetland. It is common in the Lake District, strangely absent from the mountains of north Wales, and a rarity in Ireland, growing only in the mountains of Kerry and Wicklow. It is one of the main constituents of summit heath vegetation, growing abundantly in shattered debris of mountain tops where the soil is poor and acid, and leached by constant rain. On these high tops it is frequently accompanied by Bearberry, Crowberry, and Trailing Azalea. Alpine Lady's-mantle is dwarf and neat with a tough, woody stock. The leaves are divided to the base into five to seven leaflets, and are dark green and shiny on top, and beautifully covered in silky, silvery hairs underneath. The yellow-green flowers have four sepals and no petals, and are carried in dense clusters on short stalks at the edge of the mat of leaves.

Downy Rose

Rosa villosa L.

0·4-2·0m

Downy Rose is one of the most handsome of the British roses. It usually grows as an erect bush in woods and hedges, and on mountains from Yorkshire northwards throughout Scotland; it is abundant in some Highland glens. It is the northern counterpart of the more southerly Dog Rose. Elsewhere in Britain it occurs rarely. The leaves are an attractive bluish green, with a double-toothed edge, and bear downy hairs on both surfaces, from which the plant is named. The prickles on the stem are straight. The flowers of Downy Rose vary from pink to deep red and, in certain areas, as around Lochinver in Sutherland, the deep-pink form predominates. The hips of Downy Rose are large and bristly. After the petals have fallen, the sepals remain erect on top of the hip as it ripens. Two other roses with downy leaves similar to *R. villosa* have also been called Downy Rose. *R. tomentosa,* which occurs more frequently in southern England, can be differentiated by the sepals, which turn down and are then shed as the hip ripens, and by its curved prickles. *R. sherardii* grows on more acid soils and is locally common in the north in the same areas as *R. villosa*. However, like *R. tomentosa,* it sheds its sepals as the fruit ripens.

Grass-of-Parnassus

Parnassia palustris L.

80–300mm

Grass-of-Parnassus is widespread and not uncommon in fens and bogs in Suffolk and Norfolk, and then occurs with increasing frequency in moist areas of dune slacks, fens, and wet moorland from Yorkshire and Lancashire northwards, especially on the west side of Scotland. It is absent from southern England, most of Wales except for a few places in the extreme north and in Anglesey, and from the south-west of Ireland. In bogs near Sheringham, Norfolk, Grass-of-Parnassus is plentiful, the tall flowers like big, white buttercups, looking magnificent among the equally abundant Marsh Helleborines *(Epipactis palustris)*. It is a common plant in the damp areas of the machair in the Hebrides, and is well worth looking for in the grassy patches which run down among the rocks behind so many of the beaches on the west coast of Scotland. The flowers of Grass-of-Parnassus, although outwardly like white buttercups, are beautiful and show the most delicate symmetry. The five white petals bear a pattern of deep translucent veins. Around the periphery are five long stamens, and between them are clusters of tiny, yellow filamentous scales – indeed a flower for the gods.

Large-flowered Evening-primrose

Oenothera erythrosepala Borbás

0·45–1·5m

All the evening-primroses which grow in Britain come from plants which were originally introduced from America into gardens, but have escaped and become naturalized. Large-flowered Evening-primrose is the most widespread and common species, growing in England and Wales particularly in the south from Hampshire to Cornwall. It has been found as a rare casual in south-east Scotland. It grows in waste places, particularly on sunny railway banks and in sand dunes by the sea. The biennial plants are tall and stately, up to 1·5 metres high, with a large rosette of long, pointed leaves, which are slightly toothed and crinkled along the edges. The stem and the long, red-striped sepals are covered with hairs which have bulbous red bases, by which the species can readily be recognized. The large yellow flowers, with red sepals folded back beneath them, open in the evening and are faintly scented; the lobed stigma is longer than the surrounding stamens. Common Evening-primrose *(O. biennis)* grows more in the east of England, very rarely in south-east Scotland and in eastern Ireland. It is distinguished by its smaller flowers, green sepals, and absence of red hairs; the stigma is no longer than the stamens.

Sea-holly

Eryngium maritimum L.

Although Sea-holly is a member of the Umbellifer family it bears little outward resemblance to a typical umbellifer such as Cow Parsley. The powder-blue flower head is compact and dome shaped, so that the familiar umbel shape is lost; each floret has a spiny, three-pronged bract, and five minute, linear, blue petals and sepals, from which a prominent tuft of five stamens emerges. The bracts that form a collar around the flower head, and the rounded, long-stalked leaves below, are blue-green, with a fine white border, white veins, and strong spines. Sea-holly is locally frequent by the sea in sandy places and occasionally in shingle, where it creeps and spreads widely. It is found all around the south coast of England as far north as Flamborough Head on the east and Barra in the Outer Hebrides on the west. It also occurs on the Isle of Man and all around the coast of Ireland. Sea Bindweed *(Calystegia soldanella),* with its large pink-and-white-striped trumpets, flourishes in the same coastal sites as Sea-holly, with bright-yellow Sea Spurge *(Euphorbia paralias)* and, especially in the south and west of England, the stately Tree-mallow *(Lavatera arborea)*.

80-600mm

Scots Lovage

Ligusticum scoticum L.

Scots Lovage is very much a Scottish speciality although, in Ireland, it grows on the north-east coast and in one island site in the west, off Connemara. Apart from that, it grows all round the coast of Scotland, from the Solway in the west to Berwick in the east, throughout the western isles, Orkney, and Shetland. Scots Lovage always grows close to the sea, on shingle, in grassy places, particularly on cliffs and clefts between the rocks, where it will form extensive clumps right down to the high-tide line. It is not a common plant, and I first found it growing abundantly in a typical site, in west-facing rock clefts on the bird cliffs of Ceann a Mhara on Tiree in the Hebrides. It is a neat, stout plant, with purplish, ribbed stems, and tough, glossy, three-lobed leaves, each lobe being divided again into three. When the leaves are crushed they emit a pleasant fresh smell of celery. The umbels of white flowers are flat topped and densely packed, the head of green fruits retaining this shape. Scots Lovage evokes for me pleasant associations of sun-warmed rocks above the sea, seals, and seabirds.

150-700mm

Cornish Heath

Erica vagans L.

Cornish Heath is restricted in England entirely to Cornwall, where it still grows abundantly on the acid heathlands of the Lizard peninsula, especially on Goonhilly Downs, and on the north coast of Cornwall also. It has suffered depradation from selfish people who dig it up and attempt to grow it in their gardens – it should be left alone for all to enjoy. A single patch of Cornish Heath has also been found at Belcoo, in County Fermanagh, Ireland, but the status there is open to some doubt, although in Europe it grows very locally in France from Brittany southwards, in the western Alps, and in the Pyrenees at heights up to 1800 metres. Cornish Heath has a woody base and a densely branched, moderately tall habit. The leaves are smooth, with down-rolled margins and pale undersides, and grow in whorls up the stem and above the dense, leafy flower spike, so that clumps of Cornish Heath have a fresh green look even in full bloom. The long-stalked flowers are very pretty, pale-pink, open-mouthed bells, from which the long pink stamens protrude, the anthers forming a dark-brown ring well out from the petals.

150-800mm

Scottish Primrose

Primula scotica Hook.

A long journey is necessary for most of us if we are to see Scottish Primrose, because this gem among primulas is restricted to short, cliff-top turf and calcareous dunes in the extreme north of Scotland and in Orkney. The northern coastline from Durness in Sutherland eastwards to Scrabster in Caithness is fascinating and rich in unusual flowers. Where the limestone rock forms headlands, Scottish Primrose grows all along the edge of the cliffs where the turf is very short, sometimes with Purple Oxytropis *(Oxytropis halleri)*. Near Durness, at a site where boggy ground lies over the limestone rock, it grows with Scottish Asphodel *(Tofieldia pusilla)* and, near Tongue, Scottish Primrose flowers in damp flushes where springs filter through the calcareous sand. In Orkney it is particularly fine, and grows in cliff-top turf on many of the islands. Scottish Primrose has a neat rosette of blunt leaves, that are grey underneath; leaves and stem are mealy. The flower stem is short, and the purplish flowers, which have attractive pale-yellow eyes, are borne in an erect cluster. It has a double flowering season, from May to June and from July to August, the Orkney plants being at their best in May.

20-100mm

Alpine Forget-me-not *Myosotis alpestris* F W Schmidt

70–150mm
★6–10mm

Alpine Forget-me-not was first described by George Don in the late eighteenth century, when he found it on Ben Lawers in Perth. In Scotland it is known only from Ben Lawers and a neighbouring mountain , where it grows on rocky ledges high up. It is now extinct on Caenlochan in Angus, but a letter of 1885 in the *Journal of Botany* revealed that Alpine Forget-me-not, with seed of several other European alpines, had been sown there deliberately. It grows in a number of sites in Teesdale near Mickle Fell high up in grass in a loamy soil which overlies the limestone rock. The areas are flushed with calcareous water from further up the slopes and, with the Alpine Forget-me-not, grow Spring Gentian, Mossy Saxifrage, Bird's-eye Primrose, and masses of Mountain Pansy. Alpine Forget-me-not is neat growing. It has spoon-shaped leaves that are hairy underneath and the lower ones have long stalks. The hairy stem is almost unbranched, and carries a cluster of flowers as large as those of Wood Forget-me-not *(M. sylvestris)*, and of truly Mediterranean sky blue. They are fragrant, especially in the evening.

Dodder *Cuscuta epithymum* (L.) L.

rootless parasite ★2mm

The dodders comprise a fascinating genus, related to the bindweeds (Convolvulaceae). They are wholy parasitic and devoid of chlorophyll, attaching to the stems of the host plants with suckers. The leaves are reduced to tiny scales so that the plant consists of a spider's web of fine red climbing stems, which twist anticlockwise around the host and drape themselves over it, dotted with tight spherical clusters of pink or white flowers. The spreading petals are pointed and the stamens and style project from the mouth of the flower. The old word 'dodder' means to quiver or shake, and may relate to the flowers, which will quiver when the host plant is touched. It has also been used to describe some of the bindweeds and Corn Spurrey *(Spergula arvensis)*. Dodder is widely distributed in the south of England from Cornwall to north Norfolk, in south Wales, and then rarely in Lincolnshire, Durham, the Isle of Man, and a few coastal sites in Ireland. On acid heaths and moors it is not uncommon as a parasite on Heather and gorses *(Ulex* species); on calcareous soils it parasitizes many hosts, including clovers *(Trifolium* species), thyme *(Thymus* species), and bedstraws *(Galium* species).

Hoary Mullein *Verbascum pulverulentum* Vill.

Hoary Mullein is native only in Suffolk, Norfolk, and rarely in north Essex, where it grows in dry grassy places overlying calcareous soils, such as roadsides and particularly in abandoned quarries. It is a tall, stately plant up to 1·5 metres in height, forming a large rosette of grey-green, woolly leaves in the first year, and flowering in the second year. All surfaces of the leaves and stem are thickly covered with mealy white wool, which is easily rubbed off, and which gives the plant its name. The tall, rounded stem is branched at the top, with long, clustered spikes of flowers interspersed with small leaves. The flowers are flat and pale yellow; the filaments of the five stamens are all thickly covered in white hairs. The common Great Mullein *(V. thapsus)* is also tall, but less woolly, with winged stems which are less branched. Three of the five stamen filaments are hairy like those of Hoary Mullein, but the lower two are smooth. It is a widespread plant in England and Wales, but rare elsewhere. A number of other mulleins grow in Britain, most of them rare and found only in the south. Hybridization between the various species is not infrequent, and many are grown in gardens, from which they escape and establish locally.

0·6–1·2m

Common Toadflax *Linaria vulgaris* Mill.

The toadflaxes resemble garden antirrhinums in general shape, but the flowers have a spur at the base of the lower lip of the two-lipped flower. Common Toadflax is a common perennial of grassy and waste places, railway banks, and the edges of cornfields; it is especially abundant on the chalk and frequent elsewhere. It grows throughout England, Wales, and south Scotland, less commonly in north Scotland and Ireland, and is absent from Orkney and Shetland. The plants have a creeping rhizome and numerous erect stems with masses of smooth bluish-green leaves. The bright-yellow flowers have a three-lobed lower lip, which has an orange swelling at its base, called the palate. The spur of Common Toadflax is long, straight, and points downwards parallel to the stem. The erect flower heads are densely packed, often with more than twenty flowers, and are most attractive. Two other yellow-flowered toadflaxes grow in Britain. Sand Toadflax *(L. arenaria)* is a low, sticky annual, with short-spurred flowers, growing at Braunton Burrows in North Devon. Prostrate Toadflax *(L. supinum)* is an introduced species long established in Cornwall and South Devon. It has small yellow-and-orange flowers in rounded clusters.

300-800mm

Round-leaved Fluellen

Kickxia spuria (L.) Dumort.

The name 'Fluellen' is a corruption of the Welsh; 'Llewellyn's herbs' was originally used for certain members of the genus *Veronica* but, in 1756, Sir J Hill misapplied it, and the name has persisted. Round-leaved Fluellen grows mostly in the south and east of England, but is known as far north as Lincolnshire, in the south-west of England, and in south Wales. It is a decreasing weed of cultivated ground, commonest in cornfields on chalk and limestone. It is a weak, sprawling plant, which will occasionally clamber a little way up the corn stalks, with blunt oval leaves arranged spirally up the stem; both stem and leaves are softly hairy. The solitary flowers are carried on long stalks from the bases of the leaves and, despite their small size, they are strikingly handsome. The shape is similar to that of the toadflaxes *(Linaria* species), with a short, curved spur but, in this case, the entire lower lip of the flower is pale yellow, and the two-lobed upper lip is dark maroon. Sharp-leaved Fluellen *(K. elatine)* has become rarer in recent years. It often grows with *K. spuria,* from which it can be distinguished by its slender habit, halberd-shaped leaves, and smaller, paler flowers with a straight spur.

50-250mm

Crested Cow-wheat

Melampyrum cristatum L.

Four species of cow-wheat grow in Britain; all of them are semi-parasitic on grasses, from which they derive part of their nourishment. Crested Cow-wheat grows only in East Anglia, where it is a very local plant on the edges of woods on calcareous soil, especially around Cambridge and in north-east Bedfordshire. It is a stout, branched plant with a dense flower head which is four-angled, unlike the commoner species which have flowers borne in pairs up the stem. The flowers are yellow variegated with purple, tubular and two-lipped, and almost hidden at the base of the extraordinarily shaped bracts. The bracts curve outwards and downwards, and have prominent, sharp, rosy purple teeth along either edge. The ends of the lower bracts are green, untoothed, and elongated so that they hang down below the flower head. Common Cow-wheat *(M. pratense)* grows in woods and on heaths where the humus is acid, in all parts of Britain, while the similar Small Cow-wheat *(M. sylvaticum)* with smaller orange flowers, grows in woods in northern England and Scotland. The fourth species, Field Cow-wheat *(M. arvense),* has rosy pink bracts with long teeth erect above the flowers. It is now very rare in cornfields in eastern England.

150-400mm

Yarrow Broomrape

Orobanche purpurea Jacq.

With the exception of Common Broomrape, all the other broomrapes which occur in Britain are now rare and have decreased steadily during the last few years. Yarrow Broomrape has ben recorded in the past in Pembrokeshire, Somerset, Dorset, and the Isle of Wight but, in recent years, it has only been known to flower in east Kent and on the coast of north Norfolk. Like the other broomrapes, Yarrow Broomrape is devoid of chlorophyll and, because photosynthesis is impossible, the plants are wholly parasitic upon Yarrow *(Achillea millefolia)* and rarely other composites. The underground tubers are attached to the roots of the host and, from them, the bluish flower stem arises. The flowers are also blue, veined with darker purplish blue, which readily distinguishes them from any other broomrape growing in Britain. Even the shrivelled flowers can be distinguished because this species has three bracts below the toothed calyx, while all the others have a single, pointed bract. In the Channel Islands Yarrow Broomrape is much less rare, and it grows in quantity in the north of the island of Alderney as well as in other islands.

150-300mm

Water Mint

Mentha aquatica L.

Water Mint is the commonest and most widespread of all the mints, growing throughout Britain and Ireland, less commonly in the Highlands of Scotland. It grows in wet places by streams and lochs, in ditches, and in marshes, where the creeping rhizomes can spread in the mud. It avoids anywhere that is well drained or dry. Water Mint is a very hairy plant, with a stout stem which is often purple tinged. The stalked leaves are blunt ended, toothed, and hairy like the stem. The flowers are borne in dense, rounded whorls at the top of the stem, with a second whorl often present at the leaf junction below the main flower mass; other rounded whorls cap the ends of the lateral stems. Like all other mints, the flowers have a hairy calyx with five uneven teeth, and a lilac-blue corolla with four unequal petal-like lobes. Both the style and the four stamens protrude from the mouth of the flower. When crushed, the leaves and flowers release a strong minty smell. All the wild mints hybridize readily among themselves and with garden forms which escape. Water Mint hybridizes readily with Corn Mint *(M. arvensis)*, which is the other common mint, a sprawling plant with no terminal flower mass.

150-600mm

Large-flowered Hemp-nettle

Galeopsis speciosa Mill.

Large-flowered Hemp-nettle has the most splendidly coloured flowers of all the Labiatae, rivalling the pink-and-white blooms of Bastard Balm *(Melittis melissophyllum)* in size. It is an annual weed of cultivated and disturbed ground, particularly where the soil is peaty, with a preference for turnip and potato fields. Like so many flowes of arable ground, it is nowadays much reduced both in range and quantity, but still grows in Cambridgeshire and Huntingdon, central Wales, Lancashire and Yorkshire, and particularly in south and south-west Scotland. It is a rare plant in Ireland except in Donegal and Derry. Large-flowered Hemp-nettle is robust, and has a square, branched stem and stalked, oval leaves with toothed edges and pointed tips. The richly coloured flowers are crowded in dense whorls at the top of the stem. The corolla tube is straight, twice the length of the hairy, five-toothed calyx, and pale yellow. The upper lip of the flower is pale yellow and hairy, but the lower lip is deep yellow at the throat, with a square centre lobe bearing a striking rosy purple blotch. Like all hemp-nettles, the lower lip has two small humps at the throat.

0·15–1m

Wood Sage

Teucrium scorodonia L.

Wood Sage grows commonly in most parts of Britain, but is found less frequently in the east Midlands, inland East Anglia, and south-west Ireland. It does not occur in Shetland. It grows in a considerable range of different habitats including woods, hedgebanks, acid heaths, and dry stony places, tending to avoid strongly calcareous soils and wet ground. In the north-west it can be found from sea-level in dunes and on shingles right up into the mountains. Wood Sage has a creeping rhizome and forms substantial clumps, with simple barren stems and branched flowering stems. It is rather hairy, and the paired, wrinkled leaves look like the leaves of garden sage. The flowers are borne in opposite pairs up the stem, each with a small pointed bract underneath. Wood Sage is the only labiate to have greenish-yellow flowers which have no upper lip. The lower lip is large, with two pairs of lateral lobes and a big terminal lobe which is folded down. The corolla is incomplete, so that the maroon-coloured stamens protrude from the flower. Wood Sage is of no value as a culinary herb, but a tea made from the dried leaves has been used as a preventive of rheumatism.

150–400mm

Harebell

Campanula rotundifolia L.

Known as the Harebell in England and Wales, *C. rotundifolia* is better known in Scotland as the Bluebell, and certainly in the north and west its flowers are larger and of a finer blue. It is a common plant of dry grassland, especially on chalk and limestone, growing with equal facility in heathland. In Wales and much of the north country it grows luxuriantly on sunny banks and in clefts of dry stone walls. It is uncommon in the south-west of England west of Somerset, where it is a common plant on the chalk and limestone of the north of the county. In Ross and Sutherland it is less common inland, but flourishes by the sea in the areas of machair. In Ireland it is a common plant in the north and north-west but rare elsewhere. Harebell is perennial, with creeping underground stolons, and rounded, stalked lower leaves which give the plant its scientific name, *'rotundifolia'*. The upper leaves are narrow, and the elegant flowers are usually solitary, nodding on their thin stalks. When the petals fall, the long pointed calyx teeth stick out like the spokes of a wheel. White-flowered plants are fairly frequent.

150-400mm

Venus's-looking-glass

Legousia hybrida (L.) Delarb.

Venus's-looking-glass grows in eastern and south-eastern England from east Yorkshire to Dorset, especially on the chalk soil extending from Cambridgeshire to Gloucestershire. It is an annual plant of arable ground, especially cornfields and, even in the areas where it is likely to be seen, it has become less common. It grows as far west as Bath and Somerton in Somerset, and has recently been found in the north in Cumbria. It is a branched, erect plant with pale-green, crinkled leaves, and is roughly bristly all over. The ovaries are enormously elongated, sharply three angled, and topped by the insignificant mauvish-purple flowers. The sepals are stiffly bristly and twice as long as the petals; both petals and sepals sit flat on top of the ovary, like a pair of superimposed cogwheels. The flowers close up in dull weather, making the plants difficult to see and, for this reason, they are often overlooked. Gerard first mentions the name Venus's-looking-glass in 1597, translating it from the German *Frauenspiegel,* a name still in current use. The name is said to derive from the seeds, which are oval, flat, and highly polished, but would make a very small mirror for Venus to admire herself in!

50-300mm

Round-headed Rampion

Phyteuma tenerum R Schulz

Round-headed Rampion flowers throughout July on the chalk downland of Wiltshire, Hampshire, Surrey, and Sussex. The long-stalked lower leaves are broadly lanceolate in shape, often hidden in the grass, and the upper leaves are narrow and unstalked. The striking dark-blue flower head is spherical, formed of a mass of tubular florets which curl over towards the centre. Each floret has five linear petals which form a sheath around the style and stamens. As the floret matures, the style, with its three hooked stigma lobes, pushes out through the end of the petal tube; then the petals peel back, with the tips still united around the style. The colour and shape of the flower head quickly distinguish Round-headed Rampion from the mauve flowers of Field Scabious and the smaller, paler flowers of Small Scabious *(Scabiosa columbaria)*. All three can grow in such profusion together that they colour the downland slopes. Where the turf is very short, as it is on the cliff tops by the sea in Sussex, Round-headed Rampion may be only a few centimetres high. Spiked Rampion *(P. spicatum)* is the only other rampion to grow in Britain. It is taller, with creamy, pointed flower spikes and grows in a very few places in east Sussex.

70-300mm

Sheep's-bit

Jasione montana L.

Sheep's-bit is a small perennial with dense heads of flowers like blue pincushions. They have a slight resemblance to scabious flowers, but each floret is open, small, and of the same shape; the stamens are not separate but are joined at the base. The leaves of Sheep's-bit are untoothed and form a rosette which withers as the plant flowers; the stem leaves are narrower and unstalked, the upper part of the flower stem being leafless. Sheep's-bit cannot thrive in lime-rich soils and grows in acid soils on heaths, mountains, and sea cliffs. It has a patchy distribution which is mainly western, and grows abundantly where it does occur. It grows on the heaths of Hampshire and Dorset, and is widespread in the south-west from south Somerset westwards. It is a common plant in Wales, Cumbria, and south-west Scotland, in west Scotland from Kintyre to Fort William and, although it is absent from most of northern Scotland, it is common in Shetland. In Ireland it is common around the coast, especially in the south-west. Sheep's-bit is one of the delightful features of rocky roadside banks in the west and south-west throughout late summer, usually with powder-blue flowers, but pink- or white-flowered forms are not uncommon.

50-300mm

Teasel

Dipsacus fullonum L.

Teasel is a tough biennial up to 2 metres high, the dead plant with its dried flower heads standing up gauntly throughout the winter. It is common in southern England south of a line from the Humber to the Bristol Channel, growing especially well in waste places, open woods, and riverbanks on chalk and clay soils. It is uncommon in north Wales and south-east Scotland and rare in Ireland. In the first year it forms a rosette of large leaves covered with pimple-based prickles, these leaves dying back in the second year as the plant sends up its furrowed prickly stem. The paired leaves are long and pointed, cupped around the stem, with a stout midrib roughly spined underneath, the whole leaf having a parchment-like texture with prominent veins. The bracts below the big egg-shaped flower heads are erect and curve up level with the tops. The flowers open in a band around the prickly flower head, each deep lilac floret having a four-lobed corolla deeply hidden by a long spiny bract, so that the whole head resembles a prickly, lilac honeycomb. Nectar is produced, and butterflies are frequent visitors to Teasel.

Common Cudweed

Filago vulgaris Lam.

The ten species of cudweed which flower in Britain are divided between two genera. Those belonging to the genus *Filago* have the flower heads in dense, rounded clusters at the tips of the stem branches or in the angles between them. Woolly outer bracts almost conceal the flowers, and the inner bracts are chaffy. The members of the genus *Gnaphalium* have separately stalked flower heads in a spike or umbel, each flower head being slightly conical, with a slim collar of chaffy edged bracts at the base. Common Cudweed grows in dry places on acid, sandy soils, on heaths and in dunes, mainly in eastern and south-eastern England and, in most areas, it is a declining species. It is a small plant, with grey-green, wavy edged leaves spiralling up the stem, which branches at the base and below the flower heads. These are typical of the genus *Filago,* having rounded clusters of up to forty florets sunk in a mass of cottony down. The clusters look yellow, because the outer bracts have yellow, bristle-ended tips, but the tiny tubular florets are white with red tips. The flowers of Common Cudweed are never overtopped by the leaves, unlike those of the rare Broad-leaved Cudweed *(F. pyramidata)* and Red-tipped Cudweed *(F. lutescens)*.

Sea Aster

Aster tripolium L.

Sea Aster is closely related to the garden Michaelmas Daisy so that, when it comes into flower along the edges of the saltings around the coast, it seems, like the garden flower, to herald the beginning of autumn. Sea Aster is fairly common on saltmarshes, tidal estuaries, and sea cliffs and all the coasts of Britain and Ireland, except Shetland, growing inland as far as the brackish water reaches up the rivers, with Thrift, Annual Sea-blite *(Sueda maritima),* and Common Sea-lavender *(Limonium vulgare).* It has a thick rhizome and erect, branched stems with dark-green, pointed leaves which are markedly fleshy. The attractive flowers are carried in flat-topped clusters, and resemble Michaelmas Daisies with fewer petals. The purple ray-florets around the edge of the flower are rather untidy, with gaps between them, and the centre of the flower is filled with golden-yellow, tubular, disc-florets. A variety lacking the purple ray-florets, var. *discoideus,* is fairly common in the south of England, and grows with the normal form. The seeds are achenes, and have a pappus of white, silky hairs.

Scentless Mayweed

Tripleurospermum maritimum (L.) Koch

Scentless Mayweed grows throughout Britain, rarely in the Highlands of Scotland, and is mostly coastal in Ireland. It occurs in two distinct forms which grow in different types of habitat. The inland form, *T. m.* ssp *inodorum,* has fine, thread-like leaf segments and is annual, rarely perennial, as a weed of arable and waste land. It grows on all types of soil, and is one of the first plants to colonize the edges of muddy gateways to arable fields. The seaside form, *T. m.* ssp *maritimum,* is much longer lived, usually perennial, and has blunt, fleshy leaf segments. It is a common plant of dunes, shingle beaches, and especially on the edges of sea cliffs in the west. As the name suggests, the plants are virtually scentless, unlike the strongly unpleasant-smelling Stinking Mayweed *(Anthemis cotula).* The sepal-like bracts are smooth, with a narrow brown margin, while those of Stinking Mayweed are downy, whitish, and marked with a green midrib. The solitary flowers are also larger, with longer petals and, unlike Corn Chamomile *(Anthemis arvensis),* there are no scales among the yellow disc-florets. In fruit the flat yellow disc in the centre of the flower becomes a solid cone.

Tansy

Tanacetum vulgaris L.

In the past Tansy has been so widely grown as a herb for its medicinal properties that it is likely that many of the plants found today have originated from those in cultivation. An oil was extracted from the flowers, or they were dried and used in the preparation of an infusion taken as a tonic. Tansy is locally common by roadsides and in waste places, usually near water and often beside rivers. It is less common in Scotland and Ireland. An impressive and decorative plant up to 1 metre tall, it has a very tough stem which may be tinged with red, and a mass of dark-green, pinnate leaves, deeply cut and acutely toothed. The leaves are smooth and pitted on both surfaces with tiny glands. The leaves and flowers are pungently aromatic. Around the base of each flower are several rows of phyllaries which have chaffy brown edges. The button-shaped flower heads are rayless, and composed of a tightly packed cushion, of tiny, yellow, tubular florets, with a sunken dimple in the centre of each flower head. Tansy is one of our native plants which would grace any herbaceous border, flowering well into November.

0·3-1·0m

Carline Thistle

Carlina vulgaris L.

This attractive thistle flowers throughout England in dry grassland, most commonly on the chalk downs of the south. Elsewhere it grows in calcareous dunes and on grassy cliffs where the rock is base rich, on both the east and west coasts of Scotland and across the centre of Ireland. Carline Thistle is biennial, forming a prickly rosette of leaves which dies away as the plant flowers in the second year, so that the flowering plant has stem leaves only. The dumpy stem is often purple flushed, and stem and leaves are slightly downy. The bristly outer bracts form a collar below the flower, while the prominent inner bracts, shiny and straw coloured, open in sunny weather around the true flower like the rays of a sun. There are no ray florets, all the florets being tubular and densely packed in the centre of the flower. The purple corolla lobes are not easy to see and, when the floret matures, the stigma protrudes, looking like a purple match with a yellow head. The dead plants of Carline Thistle persist throughout the following winter, as attractive in their dry state as they were when in flower.

100-500mm

Dwarf Thistle

Cirsium acaule Scop.

In England, Dwarf Thistle grows mainly south of a line from north-east Yorkshire to Dorset. It is a common thistle of short-grazed grassland, especially on chalk downland where it can be a considerable nuisance, and it is rare elsewhere. It is perennial, with a rosette of leaves growing flat on the ground and, by its extremely prickly nature, may be more easily felt than seen. The flowers are large, usually solitary, and stemless, so that the reddish-purple flower is set in the middle of the leaf rosette. Occasionally plants may be found in longer grass with stems up to 150 millimetres tall. The phyllaries below the flower head are all tightly appressed to form a cone shape, the outer ones bearing a short spine and the inner ones blunt; both may be flushed with purple. All the florets are tubular and reddish purple. Members of the genus *Cirsium,* which includes Spear Thistle *(C. vulgare)* and Marsh Thistle *(C. palustre),* can be separated from all the other thistles by the feathered hairs of the pappus, which is the dispersal mechanism for the seed. The pappus hairs of other thistles are simple.

*40–45mm

Melancholy Thistle

Cirsium heterophyllum (L.) Hill.

Whereas Dwarf Thistle is southern and low growing, Melancholy Thistle is tall and grows from mid-Wales and Derbyshire northwards with increasing frequency. It is rare in the Outer Hebrides and absent from Orkney and Shetland. In Ireland it is very rare, growing only in Sligo and Fermanagh. Melancholy Thistle is a plant of upland grassland, river sides, and road verges, spreading by underground stolons to form dense patches. In western Argyll I have seen a superb show of it in clearings between new forestry blocks, before the trees have grown too large and cut out the light. The soft, drooping leaves, like broad spear heads, are felted with white hairs underneath and with a few soft prickles along the leaf margins. The stem leaves are rounded at the base and half clasp the stem. The stems are usually unbranched, grooved, and cottony, with large solitary flowers on the ends of very long leafless stalks. The outer phyllaries below the flowers end in short points, and all the phyllaries are appressed and usually tinged with purple. There are no ray-florets, the long purple florets being tubular.

0·3–1·5m

Greater Knapweed *Centaurea scabiosa* L.

Greater Knapweed is a common plant of dry grassland, sunny banks and waysides on calcareous and base-rich soils, and is widespread in eastern and southern England from Cumbria southwards. In the west it grows on the coast of north and south Wales, and in Cumbria while, in Scotland, it is locally plentiful between Edinburgh and Dundee on base-rich grassland by the sea. On the north coast of Sutherland, it is an occasional plant of coastal sand dunes. In Ireland it is most common in County Clare and Galway on the limestone. Greater Knapweed is perenial, with stiff, branched stems and deeply pinnately cut leaves which lack the spines of thistles, but are rough to the touch. The flowers are outstanding; they can be up to 60 millimetres in diameter, much larger than those of the other knapweeds and cornflowers. The phyllaries below the flowers are green, with brown margins bearing comb-like bristles. All the florets are tubular, but the large barren outer ones have fingered lobes of such a size that they resemble the ray-florets of other composites. Greater Knapweed flowers are a favourite feeding place for the Dark-green Fritillary *(Argynnis aglaia)*.

230-800mm

Perennial Sow-thistle *Sonchus arvensis* L.

Sow-thistles are tall plants with hollow stems carrying milky juice. The dandelion-like flowers are borne in branched clusters at the tops of the stems. Apart from the rare Marsh Sow-thistle *(S. palustris)*, the other three species are common weeds of arable and waste land throughout Britain. Perennial Sow-thistle grows in all parts of Britain and Ireland, except for the upland areas of northern England and Scotland, in fields, by streams, and road verges. It has white underground stolons and can reach 1·5 metres in height. The leaves are shiny on the upper side and greyish below, the pinnate lobing not extending to the midrib, with soft spines on the leaf margins. The lower leaves clasp the stem with rounded, shallowly toothed lobes. The upper part of the stem and the bracts are covered with numerous yellow, sticky hairs, and the large flowers, deep yellow in colour, form a dandelion clock in fruit. Smooth Sow-thistle *(S. oleraceus)* has smaller, paler flowers and lacks the glandular hairs. The whole plant is smooth and rather grey, the basal lobes of the clasping leaves having triangular teeth. Prickly Sow-thistle *(S. asper)* has dark-green, prickly leaves; the ear-shaped basal lobes to the leaves are armoured with prickles.

0·3-1·5m

Flowering-rush *Butomus umbellatus* L.

Despite its name, Flowering-rush is not related to the Rush family (Juncaceae), but to the water-plantains (Alismataceae). It is a plant of rivers, ditches, and the sides of ponds and canals, where the water is slow moving and there is a thick muddy bottom. It has a patchy distribution throughout England and Wales; in Somerset and Sussex it grows in the low-lying reclaimed marshes, the Levels, but it is most common in canals, meres, and waterways in the Midlands. It does not grow north of Carlisle and, in Ireland, it is to be found mostly in the north-east around Lough Neagh. Flowering-rush is tall and stately with dark-green, three-angled leaves growing from a thick rhizome. The long narrow leaves are as tall as the leafless flower stem, which bears an umbel of beautiful pink flowers 20 to 30 millimetres in diameter. The flowers of the umbel are all long stalked and open in succession, each having three small outer sepals and three larger petals, all pink with darker-pink veins. The stamens are of two different types, some with a short filament and long anther, others with a long filament and a dumpy, heart-shaped anther. The petals stay on the flowers even when the purple fruits have formed.

0·45-1·5m

Bog Asphodel *Narthecium ossifragum* (L.) Huds.

The name *'ossifragum'* came from the old idea that animals which ate Bog Asphodel were particularly likely to suffer bone fractures. Poisoning can occur in sheep, but it takes the form of a skin hypersensitivity to sunlight induced by liver damage. Bog Asphodel is a common flower of bogs and wet, acid heaths in western Britain and Ireland. In the south and east it is locally common on the heaths of Surrey and Sussex, very local in Norfolk, and plentiful in the New Forest. Bog Asphodel has creeping roots and slightly curved leaves in two flat ranks, like a small iris. Both the stems and all the floral parts are orange-yellow, the six petal-like lobes of the flower persisting around the base of the large, pointed seed capsule. The six stamens are prominent, with orange anthers and woolly yellow filaments. It is a typical flower of bogs, where it is accompanied by cotton-sedge (*Eriophorum* species), Heather, Cross-leaved Heath *(Erica tetralix)*, and Bog Myrtle *(Myrica gale)*. The starry flowers of Bog Asphodel brighten even the dreariest moor and, in autumn, the whole plant takes on a beautiful peachy tint, and finally turns orange.

50-400mm

Marsh Helleborine

Epipactis palustris (L.) Crantz

Marsh Helleborine grows in marshes, wetlands, and damp dune slacks where the habitat is distinctly calcareous. Everywhere it is a decreasing species as wetlands are drained but it is still widely distributed throughout England, Wales, Ireland, and southern Scotland. In the coastal areas of East Anglia and west Wales, Marsh Helleborine still grows in profusion, and it has two very odd inland sites on chalk downland in Wiltshire and Bedfordshire. With their broad, hinged lips, the flowers bear the closest resemblance of any British orchid to hothouse orchids such as *Cymbidium*. Sepals and upper petals are marked with reddish brown; the cup-shaped base of the lip, the hypochile, bears parallel red veins, and the outer part of the lip, the epichile, has a frilled margin and a zigzag yellow bar across its base. Plants of the variety *ochroleuca* have no red-brown pigment, so that petals and sepals are yellow-green and the lip is white. Pollination is effected by male solitary wasps of the genus *Eumenes*. When the wasp visits the flower it lands on the hinged lip and, in trying to keep its balance, strikes the rostellum containing the pollen masses. These are glued on to the wasp's head, and carried to another flower.

200-600mm

Frog Orchid

Coeloglossum viride (L.) Hartm.

Frog Orchid is an uncommon orchid with a wide distribution throughout Britain and Ireland. It is most frequent on chalk downland in much of the south of England, where it favours short chalk turf in old chalkpits, on earthworks, and at the edges of tracks, but it has an erratic pattern of distribution. In the west and north it grows in calcareous sand dunes, in the machair of the Hebrides and in calcareous mountain pastures in north Wales and Scotland up to 1000 metres. It is unreliable in appearance, flowering in good numbers for a period, only to disappear for years at a time. The basal leaves are blunt and strap shaped while the stem leaves are pointed, merging into the stout, pointed bracts. The flower spike is dense, each flower having a neat hood of sepals and petals and a long, tongue-shaped lip which is usually tucked back under the ovary. This gives the appearance of a mass of little frog gargoyles with their tongues stuck out, one above the other. The flowers of lowland Frog Orchids are often tinged with brown, which contrasts with the pale-green lip, while those growing in the mountains tend to be entirely green. They have a sweet honey scent, which attracts small insects, and seed is set in a high proportion of flowers.

50-350mm

Southern Marsh-orchid

Dactylorhiza praetermissa (Druce) Soó

Southern Marsh-orchid grows throughout southern England as far north as north-east Yorkshire, but not in Scotland or Ireland. It is a robust orchid with a mass of broad, pale-green, unspotted leaves in a rosette at the base of the stout stem, and a large spike of flowers which may contain over 100 blooms. The flowers of typical Southern Marsh-orchid have a broad, spoon-shaped lip, with a pale central area marked by tiny black dots. The species shows considerable variation in colour, however, from pale lilac-mauve to dark magenta, and the degree of spotting on the lip may also vary. It has a strong tendency to hybridize with other marsh-orchids and spotted-orchids, so that confusing hybrid swarms are as frequently encountered as Southern Marsh-orchid itself, but the broad, conical spur, typical of all marsh-orchids, is present both in the species and its hybrids. Southern Marsh-orchid grows in fens, marshes, wet fields, and dune slacks, wherever the soil is calcareous, and even on the top of dry chalk hills. In recent years huge colonies of marsh-orchids have been found in a number of abandoned industrial sites, where toxic waste has been leached by the weather, to leave a damp, base-rich soil.

200–700mm

Pyramidal Orchid

Anacamptis pyramidalis (L.) Rich.

Pyramidal Orchid can readly be recognized by the densely packed, pyramid-shaped spike of pink flowers on a tall, unbranched stem. The upper leaves sheath the stem, while those of the basal rosette are broader and often shrivelled up by the time the plant flowers. The lip of each flower is usually deeply three lobed, with two erect plates angled across its base, but careful examination will reveal great variation in lip shape and size, even among plants in the same colony. Pyramidal Orchid is widely distributed in England, Wales, southern Scotland, and Ireland in dry pastures on chalk and limestone, and in calcareous sand dunes. Plants growing near the sea often have deeply coloured flowers, and rarely one will find albino flowers, which are particularly elegant. The flowers have a long straight spur, not as long as that of the earlier-flowering Fragrant Orchid and the scent, although sweet, is not strong. Pyramidal Orchids are pollinated by butterflies and moths, and set seed efficiently. The plant is well able to colonize new territory and, in north Wales and the Hebrides, there is evidence that it has spread widely during the last thirty years.

200–600mm

AUGUST

August is a good month to look at the flowers which grow in or beside water. All the balsams are in flower, Orange Balsam in the south and the massive Indian Balsam especially in the north of England, where it has invaded stream sides and canal banks. Monkeyflower makes a bright splash of yellow beside many a northern river while, in damp ditches and beside ponds, Great Willowherb and Greater Spearwort are in bloom. Brackish ditches near the south coast are the place for the soft-leaved Marsh-mallow, and, locally in central and eastern England in freshwater dykes and ponds, Sweet Flag can be found with its sword-like leaves smelling of tangerines. White Water-lily is widespread in lakes and peaty pools, especially in the north and west, where Water Lobelia will also be found in more sheltered, shallow areas. The rare Pipewort may be found in a few remote lochs in Argyll, Skye, Mull, and in western Ireland.

August is also the best time to look for the woodland helleborines, especially in the south of England. In the early part of the month, both Broad-leaved and Pendulous-flowered Helleborine come out, with Violet Helleborine flowering a few weeks later. Along woodland margins and in the hedges Elecampane rears its tall, stemmed heads like small sunflowers. In the past it was grown near coaching inns, the big woolly leaves being soaked in warm water and wrapped around horses' legs as a poultice to ease the strain of a hard day's driving; and plants still survive in some of its old sites. On the downs, Autumn Gentian and the mauve Clustered Bellflower can be locally abundant.

Heaths and moorlands are now beginning to colour up, Cross-leaved Heath is in full bloom, Heather and Bell Heather are beginning to come into flower, white-flowered Sneezewort is common on damp heaths; in the Breckland, the beautiful dark-blue Spiked Speedwell is in flower. On the high mountains of Scotland, Cyphel is not uncommon on bare areas of shattered rock debris, and two fascinating rarities, the tiny bright-blue Alpine Gentian and the dwarf Scottish Wormwood, flower on a few remote peaks. In the south-west and west of England, sandy beaches will have Sea Spurge and the candy-striped Sea Bindweed, with tall purple-flowered Tree Mallow in more sheltered spots along stone walls below the Tarmarisk bushes. On short turf above the cliffs the little Autumn Squill blooms, like a pale-purple version of the blue Spring Squill, and Rock Samphire is in flower on the cliffs of the south-west and west coasts and in Ireland.

White Water-lily

Nymphaea alba L.

White Water-lily is unmistakeable, with large, almost circular floating leaves, deeply notched and often tinged with purple. The large flowers also float on the water surface, having four sepals which are green on the other side, enclosing concentric ranks of more than twenty white petals. The numerous yellow stamens cluster in the centre of the flower, and the stigma is star-like, with many spreading lobes. The flowers of White Water-lily are fragrant, and it is not surprising that such a attractive water plant has also been widely cultivated throughout Britain. White Water-lily is native in most areas, growing in lakes, ponds, and canals, where the knobbly rhizomes can root deep in the mud. It is widespread in southern and eastern England, north Wales, and from there in abundance in lochs north to west Sutherland and the Hebrides. In Ireland it grows most commonly in the west. In the lochs of the Outer Hebrides and in western Connemara, a form with small leaves and flowers occurs. The fat seed capsule ripens and splits under water, each seed having air spaces in the outer coat to assist in dispersal.

★50-200mm

Cyphel

Cherleria sedoides L.

Cyphel is a true alpine, growing on bare mountain tops and screes, usually above 800 metres. In Britain it is found only in Scotland, where it is part of the typical alpine flora of the high peaks of the Grampians, around Ben Lawers in the Breadalbane Range, and especially on the bare summits of western Ross and Sutherland. Cyphel also grows in the Hebrides on Rhum and Eigg, and on the Storr in the north of Skye. Cyphel has a perennial woody root stock, from which grow tight rosettes of narrow, pointed, overlapping leaves. Mature plants form dense cushions of considerable size, which, in August, are starred with tiny bright-yellow flowers. The flowers are formed by the sepals only, because they lack petals, and have either prominent yellow stamens or styles, the sexes being carried on different plants. Old cushions of Cyphel are often tattered and torn by the wind and rain. The other flowers growing with Cyphel are similarly adapted to their hostile, exposed environment. Thrift and Moss Campion also grow in dense, low cushions, while Dwarf Willow and Trailing Azalea are prostrate and creeping.

★4-5mm

Marsh-mallow

Althaea officinalis L.

Marsh-mallow occurs mainly along the south coast of England, from Dorset eastwards, then up the coast of Suffolk and Norfolk, and in Lincolnshire around the Wash. In the west it grows in Somerset, by the Severn estuary and on the coast of south Wales, and rarely elsewhere. It is a very local plant of brackish marshes, ditches, and river banks near the sea, and may be plentiful where it does occur. The mallows which grow in Britain belong to the three genera, *Malva, Lavatera,* and *Althaea*. The genus *Althaea* is distinguished by having more than six bracts forming an epicalyx below the sepals, while the others have three. The whole plant of Marsh-mallow is velvety to the touch, both the stout stem and the rounded, stalked leaves, which have shallow lobes and teeth on their margins, and are often folded like a fan. Most of the flowers are carried at the top of the stem, with some in the leaf axils. The petals are well separated, with a shallow notch on the free border, and in the centre of the flower the stamens are bunched in a prominent boss. The roots of Marsh-mallow were used to make the sweet of the same name.

0·3–1·2m

Indian Balsam

Impatiens glandulifera Royle

Indian Balsam is not native to Britain, but was introduced as a garden plant in 1840 from Kashmir and the western Himalayas. By 1855 it had escaped from cultivation and, since then, it has spread widely; it is now to be found from Cornwall to Shetland, and throughout Ireland. It grows on the muddy banks of rivers and lakes, and is particularly abundant in the north of England. It is a massive plant reaching 2 metres in height, with ribbed red stems 40 millimetres in diameter. It is a highly successful colonizer and is just as at home on canal sides near Leeds and river banks around Dublin. The flowers are large and showy, varying in colour from white and pale pink to dark claret, while their extraordinary shape has earned them the name 'Policeman's Helmet'. Two of the three sepals are small, the third being helmet shaped with a small recurved spur like a tail. The upper petal is large and erect, with two small side petals and a further large pair forming the divided lip. Both the flowers and the foliage have an unpleasant, sickly sweet smell when crushed, and the seed capsules release their seeds explosively when touched.

0·3–2·0m

Great Willowherb *Epilobium hirsutum* L.

Great Willowherb, also known as 'Codlins-and-cream', is a very common plant of damp, sunny woodlands, stream banks, marshes, and roadside ditches. It grows throughout England, Wales, and south-east Scotland, and has been recorded as far north as Caithness, and on Barra and South Uist in the Outer Hebrides. It grows throughout Ireland. It is the tallest of the willowherbs, reaching 2 metres in height, with a stout, rounded stem. The pointed leaves are borne in opposite pairs, half clasping the stem, and both stem and leaves are densely downy with fine, spreading hairs. The buds and flowers are erect and carried in a cluster at the top of the stem and in the leaf axils, unlike the tall conical flower spike of the showy Rosebay Willowherb *(E. angustifolium)* which makes a blaze of colour in dry wasteland and on railways banks. The flowers of Great Willowherb are large with slightly notched, overlapping petals, carried on the end of a long straight ovary. The stamens shed their pollen before the creamy, four-lobed stigma is mature, so that the plant must rely upon cross-pollination by bees and hoverflies. The seeds have a simple-haired pappus and, when the long ovary splits open, they are speedily dispersed by the wind.

Rock Samphire *Crithmum maritimum* L.

The name 'Samphire' is a corruption of the French *herbe de Saint Pierre*. Both Rock Samphire and the unrelated Marsh Samphire or Glasswort (*Salicornia* species) can be made into a pickle, whose pungent flavour is an acquired taste, although it is very popular in some areas. Shakespeare mentions the harvesting of Rock Samphire on the cliffs of Dover in Edgar's famous speech to Gloucester in King Lear, 'Halfway down hangs one that gathers sampire – dreadful trade!' Rock Samphire grows on sea cliffs, rocks, and occasionally in sandy places near the sea, from Suffolk on the east coast of England right round the coast of Cornwall and Devon where it is not uncommon, to Anglesey and Great Orme in north Wales. Elsewhere, in Cumbria and western Scotland, it is a rarity of sea cliffs and, in Ireland, it is most frequent in the south and west. It is a stout, branched, greyish plant with crisp fleshy trifid leaves, the segments triangular in cross-section with pointed ends. Each leaf has a sheath around the stem. The flowers form a dense-yellow umbel; the flowers and leaves exude an oily substance when crushed or handled, and the pungent, cloying smell is very persistent.

Sea Spurge *Euphorbia paralias* L.

200–400mm

Sea Spurge is a very local plant of sandy seashores from Wigtownshire in south-west Scotland, around the west and southern coasts of Wales and England and up the east coast as far as the north Norfolk coast between Hunstanton and Wells. It is a stout perennial with a prostrate woody base from which arise barren and flowering stems which tend to branch only near the top. The leaves are narrow, blunt ended, and fleshy, glaucous green in colour, and closely packed up the stem. All the leaves are sessile, and have a barely perceptible midrib. By the time the cluster of yellowish-green flowers appears, the leaves of the lower half of the stem are often brown and shrivelled. The cluster of flowers at the top of the stem is neat and scarcely spreading, each floret having a small, smooth involucre below and roughly crescent-shaped glands set around the flower, each with stubby triangular points. The similar seaside Portland Spurge *(E. portlandica)* ranges from Islay in the north-west to Chichester Harbour in Sussex to the east. It is branched from the base, with narrow, pointed leaves often red tinged, and having a prominent midrib. It flowers earlier, from April onwards, in more grassy sites.

Cross-leaved Heath *Erica tetralix* L.

70–300mm

Cross-leaved Heath gets its name from the whorls of four narrow, pointed leaves, arranged so as to form a cross. It is a short, evergreen shrub with downy stems much softer than those of Heather and Bell Heather *(Erica cinerea)*, and hairy leaves with margins so tightly rolled that they almost meet underneath. The whorls of leaves are closely packed on the lower and barren stems, and become more widely spaced below the tight cluster of fat, oval flowers borne at the tip of the stem. The attractive flowers are delicately shaded from dark pink at the base to light at the mouth of the narrow bell; the stamens are completely hidden within the bell, and only the stigma protrudes. Cross-leaved Heath is pollinated by bees, but some bees reach the nectar by cutting a small hole through the base of the flower bell, and thus avoid pollinating the flower. It is widespread in acid soils on damp heather moors and in bogs up to 800 metres throughout western England and Wales, all of northern England and Scotland. It is absent from most of south-east and south-central England, except for the heaths of Surrey, Sussex, and Hampshire, and in the east in Norfolk and Suffolk.

Common Sea-lavender

Limonium vulgare Mill.

Common Sea-lavender is not related to lavender *(Lavendula* species) except in its colour, but to *Statice* species, whose dried flowers are often used for floral decoration. It grows in saltmarshes and muddy places by the sea, especially on the east coast from Suffolk to Essex, around Southampton Water and Poole Harbour, on the west coast of Wales, and around the shores of Morecambe Bay. It just reaches into Scotland on both the east and west coasts, but does not grow in Ireland. It has a woody root stock buried in the mud, from which grows a rosette of long-stalked, flat leaves, pinnately veined, with a spine at the leaf tip. The rounded flower stem branches near the top, and the flowers are borne in tight, flat-topped clusters grouped at the tips of the branches. The funnel-shaped calyx is papery in texture, and the anthers of Common Sea-lavender are yellow. Lax-flowered Sea-lavender *(L. humile)* has palmately veined leaves, angled stems branching well below the middle, and flowers with red anthers spaced out along the branches. Rock Sea-lavender *(L. binervosum)* is an uncommon flower of cliffs and saltmarshes in the south-west of England, south Wales and north Norfolk.

80-400mm

Marsh Gentian

Gentiana pneumonanthe L.

Marsh Gentian is a rare flower of wet heaths and bogs on acid soils; it has sky-blue flowers which are the largest of all our native gentians. It still grows from Dorset to east Sussex on the heathland, although it has disappeared from west Sussex since 1968, a sad story which is being repeated throughout its range. In the east it occurs in Norfolk, Lincolnshire, and Yorkshire and, in the west, in north Wales, Anglesey, and Furness in Cumbria. It is absent from Scotland and Ireland. Marsh Gentian is slender and erect, with narrow, dark-green stem leaves which are very difficult to spot unless the plant is in flower; there is no basal rosette. Most flowers are solitary, but rarely a robust stem will bear up to four blooms; the corolla is a glorious blue, each of the conjoined lobes marked by a green stripe on the outside. The trumpet is slightly inflated half-way down so that, when closed, the flower resembles a baggy umbrella. Marsh Gentian is remarkably resilient, flowering again, sometimes in abundance, in areas which have been burned off by heath fires. It may even benefit from the destruction of larger more invasive plants, such as gorse and bracken.

100-300mm

Autumn Gentian

Gentianella amarella (L.) Börner

Autumn Gentian is almost entirely restricted to chalk and limestone grassland, apart from areas in the north and west where it grows in calcareous sand dunes. It is a common plant of the downs of southern England, and on the chalk from the Chilterns to Norfolk, and across the centre of Ireland. It is a small, erect, much-branched biennial with no basal leaf rosette, and many narrow, blunt-ended leaves which are often tinged with purple. The flowering stem carries many blooms, the purple corolla tube having four or five lobes and a fringe of white hairs inside. The shape of the calyx is important in distinguishing the species of *Gentianella* growing in Britain; the five calyx lobes of Autumn Gentian are all linear in shape, two being slightly broader than the others, but of equal length. In northern Scotland, Scottish Gentian *(G. septentrionalis)* grows in sand dunes and limestone turf near the sea, from Ardnamurchan in Argyll to Sutherland, rarely elsewhere. It closely resembles Autumn Gentian but has a basal rosette of leaves, five pointed calyx lobes, and a five-lobed corolla which is distinctly reddish. Field Gentian *(G. campestris)* and the closely related annual *G. baltica* grow more commonly in the north, rarely in the south-east.

50-150mm

Sea Bindweed

Calystegia soldanella (L.) R. Br.

Sea Bindweed is the most attractive of all our native bindweeds, with large, trumpet-shaped flowers often candy striped in pink and white, and intermediate in size between the large white flowers of Hedge Bindweed *(C. sepium)* and the small Field Bindweed *(Convolvulus arvensis)* which may also have pink-and-white flowers. The bracts below the flowers are rounded and shorter than the sepals, not inflated or concealing the sepals as in some bindweeds. Unlike the others, Sea Bindweed is not a climber but grows in sand, more rarely in shingle, by the sea. It has stout, trailing stems and fleshy, kidney-shaped leaves with shiny surfaces and slightly wavy edges. Beneath the sand is a thick white rootstock, which creeps and spreads widely. Sea Bindweed grows on the coast of Britain from Flamborough Head in the east, all around the south and up the west coast as far north as Barra and South Uist in the Hebrides but appears to be decreasing in many areas. In Ireland it grows in scattered localities on all coasts. On sandy shores, such as those of north Cornwall, it is locally common with Sea-holly and Sea Spurge *(Euphorbia paralias)*.

*25-50mm

Monkeyflower

Mimulus guttatus DC.

Monkeyflower is an introduced species from the Aleutian Islands in North America, which was first grown in Britain in the early nineteenth century. It escaped from cultivation and established itself, first in Wales, then throughout much of Britain, especially in northern England, the borders, and eastern Scotland. It is absent from much of eastern England and, in Ireland, grows mainly in the north. Monkeyflower has a creeeping and rooting base, and is common beside gravelly streams and in wet marshy places. The upper part of the erect stems and the calyx are faintly downy, and the pale clasping leaves are toothed. The long-stalked flowers are large and showy, with a two-lobed upper and long, three-lobed lower lip, which has a hairy, orange-red spotted base and a swollen palate which almost blocks the throat of the flower. Blood-drop-emlets *(M. luteus)* from Chile has open-throated flowers bearing large red blotches on the lips, and a short stigma. It grows less commonly, mainly in Scotland. Musk *(M. moschatus)* is stickily hairy, with small, pale-yellow flowers. Grown originally for its scent, which inexplicably disappeared when the plant became naturalized, it is rare.

200-500mm

Spiked Speedwell

Veronica spicata L.

Spiked Speedwell is a rare and beautiful plant of dry, sandy heaths in the Breckland of East Anglia where the soil is basic, and of limestone cliffs near the sea in the west. It does not grow in Ireland. The western plant is taller and more robust, twice the size of the Breckland form. It has been known for many years on limestone rocks in the Avon Gorge near Bristol, and in north Somerset where it was relocated in 1912. In south Wales it is one of the many delightful plants of the limestone of the Gower peninsula, where it grows with Goldilocks Aster *(Crinitaria linosyris)* and Bloody Crane's-bill, an association repeated on the limestone of Great Orme in north Wales. Spiked Speedwell occurs again on limestone in the Severn Valley in east Powys and rarely on the limestone of Cumbria. It is perennial, with hairy, rooting basal shoots, and tall flower stems with serrated oval leaves borne mainly in the middle. The leafless flower spike is long and pointed, densely packed with deep-blue flowers, with prominent stamens protruding from them. After flowering, the tall stem and dried seed capsules persist overwinter.

80-600mm

Clustered Bellflower

Campanula glomerata L.

Clustered Bellflower is one of the characteristic flowers of calcareous grassland, and it is particularly common on the chalk downlands of southern England and the limestone grassland as far west as Bath in Somerset. It can be locally frequent on calcareous grassland in south Wales, and grows through Lincolnshire, Derbyshire, Yorkshire, and Cumbria to Angus in the north-east, where it is local but plentiful in sand dunes and on sea cliffs where the underlying rock is basic. It is rather a stiff, downy perennial, with creeping roots, the pointed lower leaves long stalked and serrated, the upper leaves stalkless. The flowers are deep-bluish purple, tightly clustered at the top of the stem, with a few more flowers in the axils of the alternate stem leaves. The lobes of the corolla are erect and pointed when the flowers first open, then gradually curl outwards at the tips, revealing the creamy white, three-lobed stigma. Dwarfed, virtually sessile plants with single flowers may be found fairly often in very short turf near the sea. They can be readily distinguished from gentians by their alternate leaves.

30-500mm

Water Lobelia

Lobelia dortmanna L.

Water Lobelia is widely distributed in shallow lakes and acid freshwater pools in north Wales, the Lake District, north-west Scotland, the Hebrides, Orkney, Shetland, and the west coast of Ireland. Water Lobelia has a submerged rosette of narrow, blunt-ended leaves, which have two longitudinal air spaces running through them and, from the creeping stolons, numerous satellite rosettes arise. The leafless stem, often tinged with red, rises above the water bearing a few drooping, pale-mauvey blue flowers. It grows in the shallow margins of pools, in sheltered bays and corners, where the delicate stems will not be smashed by the wind. Where Water Lobelia is found, the acid water does not support a wide variety of flowering plants, but those that do occur are interesting. They include White Water-lily, which usually occupies the deeper water, Shoreweed *(Littorella uniflora)*, and rarely Awlwort *(Subularia aquatica)*. In western Ireland, the islands of Skye, Mull, and Coll, and in Ardnamurchan in west Argyll, Water Lobelia may be accompanied by the rare Pipewort.

150-600mm

Common Fleabane

Pulicaria dysenterica (L.) Bernh.

150–600mm

The name 'fleabane' has been applied to several related plants which were dried and placed in beds to drive fleas away. Common Fleabane is widespread in the southern parts of England, Wales, and Ireland, becoming less common north of Yorkshire and Lancashire, and rarely recorded north of Stirling and Islay. It grows beside ponds and in marshland, damp ditches, and roadsides, where the flowers attract many butterflies, especially the Small Copper *(Lycaena phlaeas)* and the Small Tortoiseshell. Common Fleabane is perennial, with a creeping root stock and stiff, branched stems bearing pointed leaves which clasp the stem. The leaves are soft, wavy edged and, like the stems, densely covered in down, so that they appear whitish underneath. The attractive flowers are like golden-yellow daisies, borne rather sparsely in flat-topped clusters, with spreading yellow ray-florets twice the length of the tubular disc-florets. Small Fleabane *(P. vulgaris)* is a very rare plant, found in a handful of sites south of the Thames. It grows on the edges of ponds on heathland, and has small flowers with short, erect ray-florets, like little yellow brushes.

Sneezewort

Achillea ptarmica L.

200–600mm

In the past the leaves of Sneezewort were dried and powdered, and then used to stimulate sneezing. It is surprising that its close relative Yarrow was not used for the purpose because Yarrow is far more aromatic. Sneezewort grows commonly in marshes, damp grassland, and roadsides on acid soils. It is a very common plant in the north and west of Britain, less so in the south-east of England, where it is found on the small areas of damp heath, and is absent from much of southern Ireland. It is a far more handsome plant than Yarrow, with stiff, erect stems branched near the top, and dark-green pointed leaves finely saw toothed along their margins. The flowers of Sneezewort are relatively large, and borne in few-flowered, flat-topped clusters. The tubular disc-florets in the centre of the flower are grey, surrounded by the white ray-florets, which are broad with a three-lobed outer edge. The whole plant has a crisp, well-laundered look about it, beside which Yarrow looks distinctly scruffy. A double-flowered form of Sneezewort is grown in gardens, and sometimes escapes to establish itself locally.

Broad-leaved Helleborine

Epipactis helleborine (L.) Crantz

Broad-leaved Helleborine is widely distributed throughout Britain, as far north as Sutherland in Scotland, and throughout much of Ireland, but is rather less frequent in East Anglia and the north. In the south of England it is not uncommon in beech woods and on woodland road verges while, in the north and west, it also grows on limestone pavement and screes. The broad leaves are strongly ribbed and furrowed, and arranged spirally up the stem, which can be massive and up to 1 metre in height. The flower spike can also be large, with as many as 100 florets, which tend to face in one direction. Each floret has a lip with a cup-shaped base, the hypochile, which is dark brown and lined with sticky, shining nectar. The pointed outer half of the lip, the epichile, is pinkish, often folded under the hypochile, and bears at its base two rough, brown bosses, the caruncles. Plants growing in dark, shady places tend to be rather slender and greenish, while those growing in sunny sites may be dumpy, with red-flushed flowers. The similar, more slender Violet Helleborine *(E. purpurata),* grows frequently in clumps with violet-suffused stems and leaves, and has flowers with smooth, pink caruncles on the lip.

200-900mm

Irish Lady's-tresses

Spiranthes romanzoffiana Chamisso

Irish Lady's-tresses has its only European sites in Britain, but it is found widely across Canada and North America. It was first reported in Cork and Kerry in Ireland in 1810, and subsequently around the shores of Lough Neagh and its tributary streams. Since then, its range within the British Isles has been markedly extended to include Coll, Colonsay, and Barra in the Hebrides, with nine mainland Scottish sites in Argyll, and an extraordinary outlier in Devonshire, where it still persists. It is a plant of marshy meadows and the edges of lochs and rivers in areas which are clearly under water in winter. It also grows in drier rough pasture with rocky outcrops and Heather where, surprisingly, it occupies the drier ridges and not the damp gullies. The leaves are narrow and pointed in most of the Scottish plants, flatter and broader in some of the southern Irish plants, and there may be some variation in flower size and lip shape, but all may be included within the same species. The flowers are carried in three spiral ranks, and are covered with fine glandular hairs. The scent is strong and resembles that of Hawthorn, being particularly marked on warm days.

100-300mm

SEPTEMBER

The seaside in September is still an interesting area to search for plants, and may give further pleasure to the birdwatcher studying autumn migration on the coast. The edges of salt marshes are inhabited by a number of curious plants, adapted to cope with their harsh environment, where water loss is a major hazard. The glassworts are bushy, succulent plants, with stems and leaves reduced to fleshy segments which are brittle and easily snapped off. They change colour from green to yellow, red, and even purple as autumn progresses, making a brightly coloured zone on the edge of the saltings. On sandy coasts in the north of Scotland and East Anglia, Prickly Saltwort, another xerophyte like the glassworts, is in bloom with the lilac-flowered Sea Rocket. On shingles behind the beaches, especially in the south and east, Sticky Groundsel is in profusion occasionally with Sweet Alison, growing in tight little mounds like a miniature Candytuft; it is not uncommon on riverside shingles inland in the north. The dry grassland near the sea may be full of Hare's-foot Clover, with the delicate blue flowers of Pale Flax, particularly near the south coast of England.

Everywhere in the hedgerows blackberries are ripe. The distinction between the various *Rubus* species is a study for the expert botanist, but spare a glance at blackberries before you eat them, and it will soon be apparent from the varying size, shape, and flavour of the berries that there are many different types.

On chalk downs, garden lawns, and short turf such as tennis courts, the little Autumn Lady's-tresses, with its spike of spirally arranged white flowers, is in bloom – the last orchid of the year.

In a few woods and hedge banks in Kent and Sussex, wild Cyclamen can still be found. The flowers appear before the leaves, and are miniature editions of the cyclamen so well known as a pot plant. It is rare, probably introduced in most places where it grows, although it may be native in a few of them. Marsh Lousewort flowers now in boggy places on the moors and, everywhere on wet heaths and bogs, Bog Asphodel is turning orange. In a few heathy areas near the south coast, mainly in Hampshire and Devon, the beautiful blue Heath Lobelia is out. On heaths, moors, and mountains, Bell Heather is in full flower, turning purple huge area of upland which, for much of the year, are grey and austere. In fine September weather, heather walks in the hills are a delight to all the senses.

Sea Rocket

Cakile maritima Scop.

The old name of 'rocket' has been applied to many crucifers, both wild and cultivated. Sea Rocket grows all around the coast of Britain, less commonly in Ireland, on sandy and occasionally shingle shores above the high-tide line. In the north and west it can be locally plentiful on sheltered beaches, growing with Prickly Saltwort *(Salsola kali)* and Babington's Orache *(Atriplex glabriuscula),* but human pressure on seaside environment in the south has caused it to become rather scarce. Sea Rocket is an attractive annual, stout and branched, often low growing, with blue-green leaves and stem. It is smooth and succulent; the leaves are pinnately divided but reduced to short fleshy segments. The typical four-petalled crucifer flowers are large and showy, varying in colour from pale lilac to dark pink, and occasionally white. They are carried in elongated spikes. The stubby seed pods of Sea Rocket are waisted, with an upper, mitre-shaped half, and a smaller lower half shaped like a top, each containing one seed. The upper half often breaks off, leaving the lower half still attached to the plant.

100-450mm

Prickly Saltwort

Salsola kali L.

Prickly Saltwort is a rather uncommon plant which, like Sea Rocket, grows on the strand line around the coast of much of Britain. It is most frequent on the sandy shores of Lincolnshire, Norfolk and Suffolk, also in Hampshire, south Wales, and in the north-west around Islay and Kintyre. It is absent from Orkney and Shetland, and grows in scattered localities around the coast of Ireland. In many places it is decreasing and rare but, in the west of Sutherland near Dornoch Point, it is still locally abundant with Sea Rocket and Babington's Orache. It is a stout, somewhat woody, perennial, much branched and prostrate, with stems sometimes roughened by small prickles and marked by red stripes. The stalkless leaves are fleshy and succulent, almost semicircular in section, and bear a sharp spine at the tip. The flowers are solitary in the leaf axils, small, green and insignificant, lacking petals, and composed of five triangular, pointed perianth lobes which enclose the fruit. This is shaped like a top, containing a flat seed coiled like a snail shell.

*3-5mm

Hare's-foot Clover *Trifolium arvense* L.

Hare's-foot Clover is a neat, pretty little clover, easily identifiable by the soft flower heads, like furry bottle-brushes on long stems, which give the plant its name. It is widely distributed in England and Wales, less commonly in southern Scotland and Ireland, and is especially common in south-eastern and eastern England from Norfolk to Hampshire. It grows in dry sandy soil which is not calcareous, particularly in dunes and sandy pastures near the sea. Hare's-foot Clover is a slender and erect annual, with stem and leaves covered in soft spreading hairs. The flower head is soft to the touch, with tiny, pale-pink or white flowers completely hidden by the hairy calyx, which has long, pale-fawn teeth bearing fine soft bristles. The seed pods barely protrude from the calyx when ripe, and each contains a single, rounded, pea-like seed. In dry summers, Hare's-foot Clover thrives near the sea, on the sea walls of estuaries, and dry track-sides, and even in the shingles, making a pale-pink, hazy carpet where all else is sere and dead.

50-200mm

Black Nightshade *Solanum nigrum* L.

Black Nightshade grows widely in England, Wales, and the Isle of Man, but is absent from mainland Scotland and all the islands, except as a very rare casual, and grows rarely in western Ireland. In east and south-east England it can be very common as an annual weed of wasteland, and particularly in gardens and among crops on arable ground. It varies greatly in appearance from a floppy, sprawling plant to neat, domed bushes. The pointed leaves are smooth or slightly hairy, and do not have the two basal leaflets characteristic of Bittersweet, although the flowers look like white Bittersweet flowers and, similarly, have a prominent cone of yellow stamens. The clusters of drooping berries turn from green to black when ripe. Plants show considerable variation in berry shape and the position of the sepals; some variants have green or yellowish berries even when ripe. There is a perplexing variability in the toxicity of Black Nightshade, which contains the alkaloid solanine. I have known cattle to consume up to 5 kilograms daily for several weeks without ill effect while, at other times, the berries prove highly poisonous.

150-600mm

Marsh Lousewort *Pedicularis palustris* L.

80–600mm

Marsh Lousewort is a common plant of peaty bogs, marshes and wet pastures, and around springs on acid moors up to 900 metres. It is a characteristic plant of wet, open flushes, growing with Blinks *(Montia fontana),* Eyebright, and Common Butterwort. It is widely distributed in Scotland, Wales, south-west England and Ireland, and is absent from much of the rest of England. Marsh Lousewort is moderately tall, with a single branched stem which is often reddish, and deeply pinnately divided leaves with crinkled and toothed lobes, often bronzy in colour. The dark-pink flowers are carried in short terminal spikes, each flower with a bract equal in length, which is finely divided like the leaves. The calyx is hairy, with many ribs and two broad leafy lobes, and inflates when the fruit is ripe so that the seeds rattle within it, hence its other name 'Red-rattle'. The flowers are two lipped, the lips of equal length, the hooded upper lip bearing four teeth, and the lower lip broadly three lobed. Lousewort *(P. sylvatica)* is perennial and shorter, with a stout, five-angled calyx and paler flowers, the upper lip bearing two teeth, and the lower lip clearly three lobed.

Sticky Groundsel *Senecio viscosus* L.

150–300mm

Sticky Groundsel closely resembles Groundsel from which it can easily be distinguished by the larger flowers with ray-florets which curl back around the edge of the flower disc, and the sticky, glandular hairs which cover every part of the plant. These impart an unpleasant foetid smell to hands or clothing, which is not readily removed by washing, and give the plant its other common name of Stinking Groundsel. Sticky Groundsel is common and widely distributed in south-east England, the Midlands, and southern Scotland, but absent from much of south-west England, north-west Scotland, Orkney, and Shetland. In Ireland it is very rare in the east, and may have been introduced. It flourishes in dry, sandy places near the sea, in waste ground, and in the ballast of railway lines; it is a characteristic plant of dry shingle areas near the sea in late summer, but may also be found on peaty moorland. Heath Groundsel *(S. sylvaticus)* is more widespread throughout Britain and Ireland. It is taller than Sticky Groundsel with long-stalked heads of smaller, conical flowers, whose ray-florets are not folded back, and it is not so stickily hairy or foetid.

Meadow Saffron *Colchicum autumnale* L.

50-200mm

Meadow Saffron grows in damp meadows on basic soils, in rides and woodland borders. It has been much reduced by ploughing to eradicate it, because both leaves and flowers are poisonous to livestock, but it still grows in woods in north Somerset and the Cotswolds, and in meadows in south-east Wales, Yorkshire, and Suffolk. The leaves of Meadow Saffron are broad and shiny, growing in a stout cluster and appearing in March. They die back completely in June to a large underground corm, from which the leafless flower arises directly in autumn, hence the name 'Naked Ladies'. The flowers are usually solitary, the long, pale 'stem' being the base of the elongated perianth tube, which has six petal-like lobes at the top. Attached to each lobe is a stamen with orange anthers, and the three slender styles grow free from ground level inside the perianth tube. The fat green seed capsule forms in spring, and is enclosed by the leaves. Autumn Crocus *(Crocus nudiflorus)* is rarely naturalized in meadows in the Midlands and north-west of England. It has narrow leaves with a white midrib in summer, and a darker flower with three pale-yellow stamens, and a much branched and lobed orange style.

Autumn Lady's-tresses *Spiranthes spiralis* (L.) Chevall.

50-150mm

Autumn Lady's-tresses is a charming diminutive orchid, the last one to come into flower in the summer, and it is still to be found widely in short turf and lawns on calcareous and basic soils in England, as far north as Cumbria, and in many areas of Ireland. There are one or two parsnip-shaped tubers below the ground from which the flat rosette of bluish-green leaves forms in September and persists over winter. The rosette withers in early summer, and the flower spike emerges from its centre, so that the new rosette and flower appear to be unconnected. The flower spike is tightly wound in a left- or right-handed spiral. Each small, trumpet-shaped flower is composed of an upper half of conjoined sepals and petals, and a lower gutter-shaped lip, white edged and frilled. The flowers have a faint honey scent which attracts bees, seed being set in many flowers. The development from seed is very slow, and plants may not flower for fourteen years. During early life the small plants are heavily dependent on a mycorrhizal fungus which infects the tubers, and from which the plants derive most of their nutrients.

OCTOBER to DECEMBER

From October onwards the decreasing daylight and low temperatures spell the end of flowering activity for most plants, although a very few, like Night-flowering Catchfly and Black Nightshade, will still flower in sheltered spots and on the edges of cornfields where the combine harvester has passed by them.

St Dabeoc's Heath is a speciality of the late autumn but, to see it growing wild, a visit to the moorlands of Connemara is necessary. In the south and east of England it is worth looking in freshwater dykes and ponds for *Azolla,* the Water Fern. The plants float on the surface of the water where it is sheltered from the wind, and form a mat of tiny branched stems and overlapping oval leaves. In October these turn from grey-green to red, and are most attractive.

On a warm, sunny day an Ivy-covered tree can be a fascinating scene of activity. Many insects are attracted to the flowers for their nectar, and Ivy blossoms in October may be covered by hundreds of Small Tortoiseshell butterflies, as well as wasps and flies. The second brood of Holly Blue butterflies lay their eggs on the unopened flower buds of Ivy and, by late autumn, the larvae have burrowed into the buds, the withered unopened buds being a sign of their presence.

Daisies can be found flowering in most months of the year, even in December. Although few other plants have flowers at this time of year, many will already have formed new leaves and flower buds in anticipation of spring, leaves and buds which must be arranged and constructed for protection from the winter weather. The leaves of the plants which have overwintering rosettes are often downy, the dense cover of hairs protecting them against frost. Leaf and flower buds are enclosed in scales or membranes which prevent them from being damaged by cold and wet. Most annual plants, however, will pass the months of cold and darkness as seeds, protected within a tough seed coat.

Night-flowering Catchfly

Silene noctiflora L.

Night-flowering Catchfly is the latest of all the catchflies and campions, and can still be found flowering to the end of October at the edge of cornfields which are its favourite habitat. It is an annual weed of arable ground on light soils, particularly chalk and sand, and it is more common in the eastern half of England from Cumbria to Kent, in the Cotswolds, and around Bath in Somerset. Elsewhere it is a rare casual, recorded from Cornwall in the west to the coast of Angus in the north. In Ireland it grows very infrequently in the east, in County Down and Dublin. Night-flowering Catchfly resembles a pale, few-flowered Red Campion, but it softly hairy, with fine, sticky hairs on the stems and leaves, which have winged stalks clasping the main stem. The calyx is conical at first, woolly with sticky, glandular hairs, and long hairy calyx teeth. In fruit the calyx inflates, and the teeth fold back. The petals of Night-flowering Catchfly are pale pink on the upper surface, deeply cleft at the tip and, during the day, curl in over the centre of the flower, revealing their yellowish undersides. By night the petals spread out fully, and the flowers are fragrant.

100-600mm

St Dabeoc's Heath

Daboecia cantabrica (Huds.) C. Koch

St Dabeoc's Heath is one of several flowers which constitute the Lusitanian element of the flora of western Ireland, plants more readily associated with Portugal, Spain, and the south-west of France. It is restricted to heaths and moorlands in Connemara, the district, partly in the counties of west Galway and west Mayo, which lies west of Lough Corrib and south of the Twelve Bens. There it can be locally common; it has been planted and naturalized in a number of localities in Britain, and it is often grown in cultivation. St Dabeoc's Heath is a woody, much-branched perennial, with gland-tipped bristles on the flower stems. The leaves are elliptical and leathery, with edges rolled downwards, and the undersides white with fine hairs. The flowers, twice the size of Bell Heather, are fairly long stalked in a floppy spike, each shaped like a broad bell with the mouth narrowed and edged by four recurved lobes. The stamens are short and lie hidden within the flower. A double-flowered form of St Dabeoc's Heath has occurred naturally.

200-500mm

Ivy

Hedera helix L.

Ivy grows commonly in all parts of Britain and Ireland, although less often in the north of Scotland, Orkney, and Shetland. It is our only evergreen climber, flowering later in the year than most other plants, climbing on trees, rocks, and walls or trailing on the ground in scrub and woods, where it may form a dense carpet. In the north of Scotland it is more common on sea cliffs, where I have seen a massive plant with a 'trunk' as thick as a man's thigh, growing to a height of 20 metres. Ivy climbs by means of a multitude of adhesive roots, which glue themselves to the supporting surface. The lower leaves are five lobed, leathery, and shiny, with a pointed apex and triangular lobes, but the upper leaves below the rounded umbel of flowers are ovate, pointed, and unlobed. The flowers, which are borne high up on climbing plants, have five yellow-green petals which are tightly recurved, and five prominent, erect yellow stamens. The fruit is a black, five-celled berry, crowned by the remnants of the calyx and formed in early spring. Ivy berries have long been considered poisonous to children, but the leaves may be eaten by cattle, sheep, and goats in small quantities as a tonic.

to 20m

Daisy

Bellis perennis L.

The Daisy is one flower which must be familiar to everyone because it grows commonly in every part of Britain and Ireland, from sea-level to 1000 metres, in short grassland, particularly on lawns, beside roads, and in wasteland. The name Daisy comes from the old name 'Day's Eye', because the flowers close at night, concealing the bright-yellow disc at the centre of the flower – the sun – and open in the morning. The first mention of Daisy in England occurs around AD 1000 in the writings of the Saxon, Aelfric, and the image of the Daisy appears often in literature because of its pert neatness. It is perennial, with stout branched roots and a rosette of stalked, spoon-shaped leaves which are slightly toothed at the tip, and sparsely hairy. The phyllaries below the flowers are green and blunt ended. The white ray-florets are female, usually tinged with pink at the tips or on the outer side, while the tubular yellow florets, comprising the disc at the centre of the flower, are hermaphrodite. Perhaps many people will agree with Chaucer, who wrote: '. . . of alle the floures in the mede, Thanne love I most the floures white and rede, Swyche as men calle dayesyes in our toun.'

20-120mm

Wild Flowers Through The Year

In the following charts, all the flowers described in the *Calendar* are listed in alphabetical order, noting the page on which a description of the flower can be found. In the centre column of the chart, the months of the year during which each flower's three major phases of activity occur are shown by lines as follows:

growth visible above ground   in flower

in seed or fruit

Where the plant described is biannual or perennial, then some individuals within a community will probably be found in a vegetative state with growth visible above the ground at most seasons of the year, but some perennials may die back to an underground or underwater structure during the winter months. The main seasons for each plant's activity are printed continuously; in widely separated parts of Britain or at high altitude these seasons will not necessarily fall within the 'normal season'. Such extensions of the seasons are printed as broken lines.

In the right-hand column the habitats in which each species might be found are indicated by the following symbols:

deciduous woodland

pinewood

hedges and ditches

grassland, cultivated and disturbed ground

heath and moorland

mountains

freshwater, marshes, fens and bogs

seaside

A broad guide to the distribution of the flowers in Britain is also given. Britain is defined here as England, Scotland, Wales and Ireland.

Species	Jan Feb Mar Apr May June July Aug Sep Oct Nov Dec	Habitat & Distribution
Alexanders p63		coastal England Wales & E Ireland
Alpine Forget-me-not p126		Perth & Teesdale
Alpine Lady's-mantle p122		NW England & Scotland
Autumn Gentian p147		most of Britain
Autumn Lady's-tresses p156		England & Ireland
Barren Strawberry p45		all Britain less common N Scotland
Bearberry p83		N Scotland N & W Ireland rare N England
Bee Orchid p116		England, Wales & Ireland
Bird's-eye Primrose p84		N England
Biting Stonecrop p102		all Britain
Bittersweet p106		mainly England & Wales
Black Nightshade p154		England & Wales

Species	Jan Feb Mar Apr May June July Aug Sep Oct Nov Dec	Habitat & Distribution
Bloody Crane's-bill p120		mainly coast N & W Britain
Bluebell p69		absent Orkney & Shetland
Bog Asphodel p138		mainly W & N Britain
Bogbean p85		all Britain
Bog Pimpernel p104		mainly W Britain
Breckland Speedwell p55		rare E Anglia
Bristly Oxtongue p113		England & Wales
Broad-leaved Helleborine p151		all Britain
Bugle p67		all Britain
Burnt Orchid p92		England only
Bush Vetch p61		all Britain
Butchers-broom p47		mainly S England

Species	Jan Feb Mar Apr May June July Aug Sep Oct Nov Dec	Habitat & Distribution
Butterbur p56		all Britain except N Scotland
Carline Thistle p135		less common N Scotland
Changing Forget-me-not p86		all Britain
Clustered Bellflower p149		England & E Scotland
Colt's-foot p46		all Britain
Common Broomrape p108		S England & S Ireland
Common Butterwort p87		mainly W & NW Britain
Common Centaury p104		all Britain, rare Scotland
Common Chickweed p41		all Britain
Common Comfrey p85		all Britain, common in S
Common Cudweed p133		mainly S & E England
Common Dog-violet p59		all Britain

Species	Jan Feb Mar Apr May June July Aug Sep Oct Nov Dec	Habitat & Distribution
Common Fleabane p150		England, Wales & S Ireland
Common Fumitory p75		commoner E side Britain
Common Milkwort p76		all Britain
Common Orache p120		all Britain
Common Poppy p94		less common N & W Britain
Common Ragwort p111		all Britain
Common Restharrow p121		less common NW Scotland & Ireland
Common Scurvy-grass p58		all coasts rare in S
Common Sea-lavender p146		local E, S, & NW England & Wales
Common Spotted-orchid p117		all Britain
Common Stork's-bill p98		all Britain mainly coast in N & W
Common Toadflax p127		all England & Wales rare Scotland & Ireland

Species	Jan Feb Mar Apr May June July Aug Sep Oct Nov Dec	Habitat & Distribution
Common Twayblade p91		all Britain
Common Winter-green p103		scattered England, Scotland, & Ireland
Cornish Heath p125		S Cornwall & Co Fermanagh
Corn Marigold p112		all Britain
Cow Parsley p62		all Britain
Cowslip p63		less common N & W Britain
Crested Cow-wheat p128		rare E England
Cross-leaved Heath p145		all Britain
Crowberry p54		N & W Britain
Cyphel p142		Highland Scotland only
Daisy p159		all Britain
Dandelion p68		all Britain

Species	Jan Feb Mar Apr May June July Aug Sep Oct Nov Dec	Habitat & Distribution
Deadly Nightshade p86		mainly England
Dodder p126		mainly S England
Dog Rose p101		commoner S England
Dog's Mercury p45		England, Wales, & S Scotland
Dove's-foot Crane's-bill p78		all Britain
Downy Rose p122		N England & Scotland
Dropwort p100		England & N Wales
Dwarf Thistle p136		England mainly S & SE
Early Forget-me-not p65		all Britain mainly SE England
Early Marsh-orchid p117		all Britain
Early-purple Orchid p71		all Britain
Early Spider-orchid p70		rare S England

Species	Jan Feb Mar Apr May June July Aug Sep Oct Nov Dec	Habitat & Distribution
Eyebright p107		all Britain
Field Penny-cress p75		all Britain mainly SE England
Field Scabious p110		all Britain rare N Scotland
Flowering-rush p138		England, E Wales & N Ireland
Fragrant Orchid p115		all Britain
Fritillary p89		S & E England
Frog Orchid p139		all Britain
Garlic Mustard p59		less common Scotland & Ireland
Germander Speedwell p66		all Britain
Globeflower p73		N England, Wales, & Scotland
Grass-of-Parnassus p123		mainly N & NW Britain
Grass Vetchling p100		S & SE England

Species	Jan Feb Mar Apr May June July Aug Sep Oct Nov Dec	Habitat & Distribution
Greater Bladderwort p108		all Britain
Greater Butterfly-orchid p116		all Britain
Greater Celandine p74		mainly England & Wales
Greater Knapweed p137		mainly E & S England
Greater Stitchwort p60		all Britain
Great Willowherb p144		all Britain rare N Scotland
Green-veined Orchid p92		England Wales & Ireland
Groundsel p42		all Britain
Harebell p131		less common S & E Ireland
Hare's-foot Clover p154		all Britain, rare N & Ireland
Heath Dog-violet p51		all Britain
Hemlock p103		all Britain mainly S

Species	Jan Feb Mar Apr May June July Aug Sep Oct Nov Dec	Habitat & Distribution
Henbit Dead-nettle p67		mainly E half Britain
Herb-Paris p90		not Ireland
Herb-robert p60		all Britain
Hoary Mullein p127		E Anglia
Hoary Rock-rose p77		N England, Wales, & W Ireland
Horseshoe Vetch p79		England & Wales
Indian Balsam p143		all Britain
Irish Heath p53		W Ireland
Irish Lady's-tresses p151		Ireland, W Scotland, & SW England
Ivy p159		all Britain
Ivy-leaved Toadflax p65		all Britain
Kidney Vetch p99		all Britain

Species	Jan Feb Mar Apr May June July Aug Sep Oct Nov Dec	Habitat & Distribution
Lady Orchid p91		SE England
Lady's Smock p58		all Britain
Large-flowered Evening-primrose p123		England
Large-flowered Hemp-nettle p130		mainly N Britain
Lesser Celandine p50		all Britain
Lesser Spearwort p94		all Britain
Lily-of-the-valley p89		England & Wales, absent Ireland
Lords-and-Ladies p71		absent N Scotland
Maiden Pink p77		absent Ireland
Marsh Gentian p146		England & Wales
Marsh Helleborine p139		England, Wales, Ireland, & S Scotland
Marsh Lousewort p155		all Britain mainly N & W

Species	Jan Feb Mar Apr May June July Aug Sep Oct Nov Dec	Habitat & Distribution
Marsh-mallow p143		S England & S Ireland
Marsh-marigold p49		all Britain
Marsh Thistle p113		all Britain
Meadow Buttercup p74		all Britain
Meadow Crane's-bill p97		England, Wales, & S Scotland, rare N Ireland
Meadow Saffron p156		England & E Wales
Meadow Saxifrage p61		absent NW Scotland rare Ireland
Melancholy Thistle p136		Scotland, N England, & N Wales
Mezereon p52		rare England
Monkey-flower p148		all Britain
Moschatel p68	rarely produced	England, Wales, & S Scotland, rare Ireland
Mossy Saxifrage p81		Wales, N England, Scotland, & Ireland

Species	Jan Feb Mar Apr May June July Aug Sep Oct Nov Dec	Habitat & Distribution
Mountain Avens p80		N England, N Wales, Scotland, & W Ireland
Mountain Everlasting p111		N England, Wales, Scotland, & NW Ireland
Mountain Pansy p76		N England, Wales, Scotland, & E Ireland
Mouse-ear Hawkweed p114		all Britain
Musk Thistle p112		England, Wales, & SE Scotland
Night-flowering Catchfly p158		England & S Scotland
Opposite-leaved Golden-saxifrage p51		all Britain
Oxeye Daisy p88		all Britain
Oxlip p64		E Anglia
Pasque-flower p73		S & E England
Perennial Sow-thistle p137		all Britain
Perforate St John's-wort p119		all Britain less in N Scotland

Species	Jan Feb Mar Apr May June July Aug Sep Oct Nov Dec	Habitat & Distribution
Prickly Saltwort p153		coastal all Britain
Primrose p54		all Britain
Purple Saxifrage p62		Wales, N England, Scotland, & Ireland
Pyramidal Orchid p140		England, Wales, S Scotland, & Ireland
Ramsons p70		all Britain
Red Dead-nettle p46		all Britain
Ribbed Melilot p98		mainly SE England
Rock Samphire p144		coastal, mainly S & W
Rock Speedwell p107		Highlands
Roseroot p80		Wales, N England, Scotland, & W Ireland
Round-headed Rampion p132		S England
Round-leaved Crowfoot p44		S & W Britain, S Ireland

Species	Jan Feb Mar Apr May June July Aug Sep Oct Nov Dec	Habitat & Distribution
Round-leaved Fluellen p128		S & C England, S Wales
Round-leaved Sundew p102		mainly W Britain
St Dabeoc's Heath p158		W Ireland
Scentless Mayweed p134		all Britain
Scots Lovage p124		N Scotland & Ireland
Scottish Primrose p125		N Scotland
Sea Aster p134		coastal all Britain
Sea Beet p97		coastal all Britain rare Scotland
Sea Bindweed p147		coastal all Britain rare Scotland
Sea Campion p96		coastal all Britain & mountains
Sea-holly p124		coastal, not N & E Scotland
Sea-kale p119		coastal NW & S England

Species	Jan Feb Mar Apr May June July Aug Sep Oct Nov Dec	Habitat & Distribution
Sea Pea p121		E & S coast England
Sea Rocket p153		coastal all Britain
Seaside Pansy p95		coastal N & W Britain & Ireland
Sea Spurge p145		coastal England, Wales S Scotland & Ireland
Sheep's-bit p132		mainly W Britain & Ireland
Sheep's Sorrel p82		all Britain
Shepherd's-purse p41		all Britain
Shrubby Cinquefoil p101		N England & W Ireland
Silver-weed p79		all Britain
Skullcap p109		all Britain
Sneeze-wort p150		all Britain
Snowdrop p47		England, Wales, & Scotland

Species	Jan Feb Mar Apr May June July Aug Sep Oct Nov Dec	Habitat & Distribution
Southern Marsh-orchid p140		England & Wales
Spiked Speedwell p148		E Anglia, W England, & Wales
Spring Gentian p64		Teesdale & W Ireland
Spring Sandwort p96		N England & N Wales, rare Scotland & W Ireland
Spring Squill p69		coastal W & N Britain
Spurge-laurel p52		England & Wales
Squinancy-wort p109		S England, S Wales & SW Ireland
Sticky Groundsel p155		England, Wales, & S Scotland
Tansy p135		all Britain
Teasel p133		mainly S & E England
Thrift p83		coastal all Britain & N mountains
Toothwort p66		all Britain not extreme N & W

Species	Jan Feb Mar Apr May June July Aug Sep Oct Nov Dec	Habitat & Distribution
Trailing Azalea p82		N Scotland
Tufted Vetch p99		all Britain
Twinflower p110		NE Scotland
Venus's-looking-glass p131		S & E England
Viper's-bugloss p106		England, Wales, S Scotland, & E Ireland
Wall-flower p50		mainly S & E Britain
Wall Speedwell p55		all Britain
Water Forget-me-not p105		all Britain
Water Lobelia p149	submerged	N England, Wales, Scotland, & Ireland
Water Mint p129		all Britain
Water-plantain p114		all Britain, rare N Scotland
Water-violet p84	submerged	S & E England

Species	Jan Feb Mar Apr May June July Aug Sep Oct Nov Dec	Habitat & Distribution
Weld p95		all Britain, not NW Scotland
White Bryony p81		England
White Helleborine p90		S England
White Water-lily p142		all Britain
Wild Clary p88		mainly S England
Winter Aconite p44		mainly S & E England
Winter Heliotrope p42		all Britain mainly S
Wood Anemone p49		all Britain, rare S Ireland
Wood Sage p130		all Britain
Wood-sorrel p78		all Britain
Wood Spurge p53		S England & S Wales
Yarrow Broomrape p129		S & E England

Species	Jan Feb Mar Apr May June July Aug Sep Oct Nov Dec	Habitat & Distribution
Yellow Iris p115		all Britain
Yellow Rattle p87		all Britain
Yellow Star-of-Bethlehem p56		C & N England, S Scotland
Yellow-wort p105		England, Wales, & S Ireland

Further reading

Those marked ★ are available from BSBI Publications, Oundle Lodge, Oundle, Peterborough, PE8 5TN.

General reference

Ary, S & Gregory, M. 1960. *Oxford Book of Wild Flowers*. Oxford University Press, Oxford.

Bishop, O N. 1973. *Natural Communities*. John Murray, London.

Butcher, R W. 1961. *New Illustrated British Flora*. Leonard Hill, London.

Clapham, A R, Tutin, T G, & Warburg, E F. 1962. *Flora of the British Isles*. Cambridge University Press, Cambridge.

Clapham, A R, Tutin, T G, & Warburg, E F. 1981. *Excursion Flora of the British Isles*. Cambridge University Press, Cambridge.

Dony, J G, Rob, C M, & Perring, F H. 1980.★ *English Names of Wild Flowers*. Butterworths, Sevenoaks.

Fitter, R. 1967. *Penguin Dictionary of British Natural History*. Penguin, Harmondsworth.

Fitter, R S R, 1971. *Finding Wild Flowers*. Collins, London.

Fitter, R S R, Fitter, A, & Blamey, M. 1974. *Wild Flowers of Britain and Northern Europe*. Collins, London.

Flora Europaea (five volumes). 1964–80. Cambridge University Press, Cambridge.

Grey-Wilson, C & Blamey, M. 1979. *Alpine Flowers of Britain and Europe*. Collins, London.

Lang, D C. 1980. *Orchids of Britain*. Oxford University Press, Oxford.

McClintock, D & Fitter, R S R. 1956. *Collins Pocket Guide to Wild Flowers*. Collins, London.

Martin, W Keble. 1965. *Concise British Flora in Colour*. Ebury Press, London.

Nilsson, S & Mossberg, B. 1979. *Orchids of Northern Europe*. Penguin Nature Guides, Harmondsworth.

Pearsall, W H. 1971. *Mountains and Moorlands*. Collins, London.

Perring, F H & Walters, S M. 1983. *Atlas of the British Flora. EP Publishing, Wakefield.*

Raven, J & Walters, S M. 1971. Mountain Flowers. Collins, London.

Rose, F. 1981. *Wild Flower Key of the British Isles and North-west Europe*. Frederick Warne, London.

Stamp, L Dudley. 1946. *Britain's Structure and Scenery*. Collins, London.

Summerhayes, V S. 1951. *Wild Orchids of Britain*. Collins, London.

Williams, J G, Williams, A E, & Arlott, N. 1978. *Field Guide to the Orchids of Britain and Europe*. Collins, London.

Regional reference

Anglesey Roberts, R H. 1982. *The Flowering Plants and Ferns of Anglesey*.★

Angus Ingram, R & Noltie, H J. 1981. *The Flora of Angus*.★

Bardsey Conolly, A. 1980. 'Silverweed on Bardsey'. *Bardsey Observatory Report* 24 p42.

Bedfordshire Dony, J G. 1953. (Reprint.) *The Flora of Bedfordshire*.★
Dony, J G. 1976. *Bedfordshire Plant Atlas*.★

Breckland Trist, P J O. 1979. *An Ecological Flora of Breckland*.★

Caithness Bullard, E R. 1977. *The Wild Flowers of Caithness*.★

Cambridgeshire Ewen, E H & Prime C T. 1660 new edition 1975. *Ray's Flora of Cambridgeshire*.★

Carmarthenshire May, R F. 1969. *A List of Flowering Plants and Ferns of Carmarthenshire*.★

Cornwall Davey, F H. 1909. Supplement by Thurston, E & Vigurs, C C. 1922. (Reprint.) *The Flora of Cornwall.*★
Margetts, L J & David, R W. 1981. *A Review of the Cornish Flora.*★
County Carlow Booth, E M. 1979. *The Flora of County Carlow.*★
Cumbria Halliday, G. 1978. *Flowering Plants and Ferns of Cumbria.*★
Derbyshire Patrick, S & Hollick, K M. 1974. *Supplement to the Flora of Derbyshire.*★
Dorset Good, R D'O. *Good's Handlist of Dorset Flora.*★
Dumfries Milne-Redhead, H. 1972. *A Checklist of the Flowering Plants, Ferns and Fern Allies of Dumfries, Kirkcudbright and Wigtown.*★
Essex Jermyn, S T. 1974. *Flora of Essex.*★
Exmoor Giddens, C. 1979. *The Flowers of Exmoor.*★
Gloucestershire Riddlesdell, H J, Hedley, G W, & Price, W R. 1948. *Flora of Gloucestershire.*★
Guernsey McClintock, D. 1975. *The Wild Flowers of Guernsey.*★
Hebrides Campbell, M S. 1945. *The Flora of Uig (Lewis).* T Buncle & Co Ltd.
Currie, A. 1979. *The Vegetation of the Outer Hebrides.*★
Herefordshire Whitehead, L E. 1976. *Plants of Herefordshire.*
Hertfordshire Dony, J G. 1967. *Flora of Hertfordshire.*★
Highlands Darling, F Fraser & Boyd, J M. 1973. *Highlands and Islands.* Collins, London.
Plant Communities of the Scottish Highlands: a study of Scottish mountains, moorland and forest vegetation. Monographs of the Nature Conservancy, HMSO, London.
Isle of Wight Bevis, J, Kettell, R, & Shephard, B. 1978. *Flora of the Isle of Wight.*★
Kent Philp, E G. 1982. *Atlas of the Kent Flora.*★
Kintyre Cunningham, M C & Kenneth, A G. 1979*Flora of Kintyre.*★
Lancashire Weldon, J A & Willson, A. 1907. (Reprint.) *The Flora of West Lancashire.*★
Lincolnshire Gibbons, E J. 1975. *The Flora of Lincolnshire.*★
Lothian Martin, I H. 1934. *The Field Club Flora of the Lothians.*
Middlesex Kent, D H. 1975. *The Historical Flora of Middlesex.*★
Monmouthshire Wade, A E. 1970. *The Flora of Monmouthshire.*★
Montgomeryshire MacNair, J. 1977. *Plants of Montgomeryshire.*★
Moray Webster, M McCallum. 1978. *Flora of Moray, Nairn and East Inverness.*
Mull Jermy, A C & Crabbe, J A. 1978*The Island of Mull. A Survey of its Flora and Environment.*★
Norfolk Swann, E L. 1975. *Supplement to the Flora of Norfolk.*★
Northumberland Thompson, R. 1980. *A New Flora of the Alnwick District.*★
Orkney Bullard, E R. 1975. *A Check List of Vascular Plants.*★
Perthshire White, F R. 1898. (Reprint.) *Flora of Perthshire.*★
Ross-shire Duncan, U K. 1980. *Flora of East Ross-shire.*★
Rutland Messenger, K G. 1971. *Flora of Rutland.*★
Shetland Barkham, J P, Gear, S, Hawksworth, D L, & Messenger, K G. 1981. *The Flora of Foula.*★
Shropshire Leighton, W A. 1841. (Reprint.) *Flora of Shropshire.*★
Skye Murray, C W & Birks, H J B. 1980 *The Botanist in Skye.*★
Staffordshire Edees, E S. 1972. *Flora of Staffordshire.*★
Somerset Roe, R G B. 1981. *The Flora of Somerset.*★
Suffolk Simpson, F. 1982. *Simpson's Flora of Suffolk.*★
Sussex Hall, P C. 1980. *Sussex Plant Atlas.*★
Sutherland Kenworthy, J B. 1976. *John Anthony's Flora of Sutherland.*★
Teesdale Pigott, C D. 1956. 'The Vegetation of Upper Teesdale in the North Pennines'. *Journal of Ecology* 44 p545-86.
Warwickshire Cadbury, D A, Hawkes, J G, & Readett, R C. 1971. *A Computer-mapped Flora.*
Wiltshire Grose, J D. 1957. (Reprint.) *The Flora of Wiltshire.*★
Worcestershire Amphlett J & Rea, C. 1909. (Reprint.) *The Botany of Worcestershire.*★
Yorkshire Lees, F. A. 1888. (Reprint.) *The Flora of West Yorkshire.*★

Index

M

N

O

P

Q

R